Of Stones and Man: from the Pharaohs to the Present Day

BALKEMA - Proceedings and Monographs
in Engineering, Water and Earth Sciences

Of Stones and Man

from the Pharaohs to the Present Day

Jean Kerisel

LONDON / LEIDEN / NEW YORK / PHILADELPHIA / SINGAPORE

English translation: Philip Cockle
Final editing: Thierry Kerisel, Philip Cockle and Germaine Seijger*
Design & typesetting: Maartje Kuipers*
Printing: The Bath Press, CPI Group, Bath, United Kingdom

*Published by: Taylor & Francis / Balkema
P.O. Box 447, 2300 AK Leiden, The Netherlands
e-mail: Pub.NL@tandf.co.uk
www.balkema.nl, www.tandf.co.uk/books, www.crcpress.com

ISBN HB: 0 415 36441 8
ISBN PB: 0 415 38345 5

Printed in Great-Britain

The author would like to express his profound gratitude to his translator, Philip Cockle, who has become his friend, and to his son Thierry Kerisel whose great qualities as historian and engineer have been of immense assistance.

Table of contents

Jean Kerisel (1908-2005)

Jean Kerisel died in Paris on 22 January 2005 at the age of 96, just a short time before the publication of the English edition of this, his last book.

He was born in Brittany, a region in the extreme west of France to which he was deeply attached and which profoundly marked his personality. After attending school in his native town of St. Brieuc, he continued his studies in Paris at the *Ecole Polytechnique*, a seedbed for French scientists for the past 200 years, and then in 1933 at the *Ecole Nationale des Ponts et Chaussées*, the training-ground for high-level civil engineers. From 1933 to 1940 he engaged in the theoretical and practical studies that would help found the new discipline of soil mechanics, obtaining a doctorate in the physical sciences at the Sorbonne in Paris in 1935 for a thesis on friction of cohesionless soils and its applications to the design of foundations.

Like all the men of his generation he was caught up in the Second World War, during which he was awarded the *Croix de Guerre* for bravery. Towards the end of the war he became Director of reconstruction at the Ministry of Reconstruction (1944-1951) with the heavy responsibility of organizing the rebuilding of the major French cities. During this period he published two seminal works on geotechnical engineering, a treatise on soil mechanics and *Tables for the calculation of passive pressure, active pressure, and bearing capacity of foundations*. Both works were translated into numerous languages and frequently reprinted.

In 1952 he left the public sector to found Simecsol, a research consultancy specializing in geotechnical engineering and infrastructures while continuing to serve as Professor of Soil Mechanics at the *Ecole Nationale des Ponts et Chaussées* (1951-1969). While President of Simecsol he was involved as designer or consultant in a large number of major engineering projects, including bridges in the Ivory Coast, Nigeria, Venezuela and France, dams, tunnelling operations for the new Ligne A of the Paris underground system and for the CERN ring at Geneva, the building of ports (Sibari in Italy and Douala in Cameroon) as well as the foundations for industrial complexes, nuclear power stations and tower blocks.

As an internationally recognized member of his profession, in 1969, he was elected President of the French Committee of Soil Mechanics and Foundation Engineering, resigning in 1973 to become President of the International Society for Soil Mechanics and Foundation Engineering. In 1968-1969 he was also President of the French Society of Civil Engineers.

He has received a number of distinctions in France (Commander of the Légion d'Honneur in 1975) and other countries. In 1975 he was invited to give the 15th Rankine Lecture in London for the British Geotechnical Society on "Old structures in relation with soil conditions". He was made an Honorary Member of the Hungarian Academy of Science and a Doctor Honoris Causa of the Universities of Liège and Naples. For a whole generation of engineers, who studied under his guidance or worked with him on the design and construction of numerous infrastructures in the field of civil engineering, he was the master, adviser and friend who fostered the development and furthered the recognition of a still young scientific discipline. He kept in touch with his pupils, assistants and colleagues throughout the world and was valued and esteemed everywhere.

From 1979 Jean Kerisel's activities took a new direction. He now had more time to devote to his many interests, his friends and his family, which was steadily growing in size with the birth of grandchildren and great grandchildren. He was a tireless worker – he once told his translator, his friend Philip Cockle, that the secret of old age was to 'never down tools' – always coming up with new ideas and projects. During a professional visit to Cairo in connection with the construction of the city's underground transport system his old interest in ancient Egypt was rekindled. He was fascinated by a civilization that had thrived for some three thousand years before our era, leaving behind countless examples of its culture and technical skill and the problem of piercing their secrets. It was precisely his long experience as an illustrious civil engineer that gave his texts on ancient Egypt – and on other major constructions of the past – their special flavour. He looked at the pyramids more as an engineer than as an archaeologist, acutely aware of the technical difficulties their builders had to overcome.

To take another example, he had long been interested in the problems of the Tower of Pisa, which was leaning more and more each year and in imminent danger of collapse. Its fall would be catastrophic for the local economy, but no one wanted it to be made perfectly vertical! Many possible solutions were proposed – some mentioned in this book – which he combated vigorously, until he learned of the proposals by John Burland, Professor of Soil Mechanics at Imperial College, London. "That's the answer", he exclaimed, and immediately threw his weight behind them. John Burland's idea was eventually adopted and has proved a great success.

His passion for the history of foundation engineering, and a taste for writing, led to the publication of four earlier books: *Down to Earth – Foundations past and present: the Invisible Art of the Builder* (1987); *La Pyramide à travers les âges, Mythes et Religions* (1991); *Génie et démesure d'un pharaon: Khéops* (1996); and *The Nile and its Masters: Past, Present, Future. Source of Hope and Anger* (2001). He also wrote a biography of his intellectual master and father-in-law, Albert Caquot (1881-1976). The first edition, published in 1978, was later considerably expanded and re-published in 2001 under the title *Savant, Soldat et Bâtisseur*, which sums up well the long career of one of France's foremost engineers in the twentieth century.

Jean Kerisel remained young in spirit to the end and achieved a perfect balance in his personal, family and professional life. With his lively intelligence, broadness of mind and warm-heartedness, he was always ready to listen to others yet pursued his goals with unfailing courage and determination. All who knew him were attracted by his simplicity, openness and directness, though his natural consideration for other people was never allowed to deform his judgment.

Following the death of his wife in 1998 after sixty-six years of happy marriage, he continued to maintain a welcoming home for his family and friends. Despite, too, the *infinite melancholy of old age,* my father, Jean Kerisel, retained until the last his lucidity, youthful spirit and love of life. Among the favours granted to this exceptional man the most precious was not to experience a twilight at the end of his long existence.

Thierry Kerisel

By the same author:

Tables de butée, de poussée et de force portante des fondations and *Tables for the calculation of passive pressure, active pressure and bearing capacity of foundations*, co-author with Albert Caquot, (1948), Gauthier-Villars, Paris. Re-published in 1973, bilingual edition French-English, co-author with Albert Caquot and Elie Absi, Gauthier-Villars/Bordas, Paris, reprinted in 1990 and 2003.

Traité de Mécanique des Sols, co-author with Albert Caquot. Re-published in 1949, 1956 and 1966, Gauthier-Villars, Paris, translated into German (1967), Spanish (1969), Rumanian (1968) and Japanese (1975).

Glissements de terrains, Abaques, Dunod, Paris (1967).

Albert Caquot, Créateur et Précurseur, Eyrolles, Paris (1978).

Down to Earth, Foundations Past and Present, the Invisible Art of the Builder, Balkema, Rotterdam (1987, reprinted in 1991).

La Pyramide à travers les âges. Mythes et religions, Presses de l'ENPC, Paris (1991).

Génie et démesure d'un pharaon: Khéops, Stock, Paris (1996, reprinted in 2001).

Le Nil: l'espoir et la colère. De la sagesse à la démesure, Presses de l'ENPC, Paris (1999).

The Nile and its Masters; Past, Present, Future, Source of Hope and Anger, Balkema, Lisse (2001).

Albert Caquot (1881-1976), Savant, soldat et bâtisseur, Presses de l'ENPC, Paris (2001)

Pierres et Hommes, des Pharaons à nos jours, Presses de l'ENPC, Paris (2004)

Of Stones and Man: from the Pharaohs to the present day, Taylor & Francis/Balkema, Leiden (2005).

Introduction

"*Mens agitat molem*", wrote Virgil in the Aeneid (Book VI, line 727), "*Mind enlivens the whole mass*". For Virgil, a spiritual principle animated the world; today, seen more narrowly, his words remind us that man's intelligence has given him a powerful hold over matter. Is this ascendancy good or bad? Is it unbounded or does it have limits? That is the theme of this book.

The poet Lucan lived in Rome during the first century of our era. He became interested in the early days of the city, then the capital of the ancient world. He had been told about the buildings that used to overlook the Aventine within the *Roma Quadrata* – but by his time they had been reduced to dust: "*etiam periere ruinae*", he wrote, "*Even the ruins perish*". He was saddened by the passage of time and its destructive powers. Soon afterwards - irony of fate - the Emperor Nero punished his involvement in an obscure plot by commanding him to take his own life.

Today the Aventine is still a bare mound. Its memories of past civilizations have been utterly lost and already it is surrounded by the remains of deserted imperial palaces, isolated porticos and peristyles eaten away by pollution. That is one view of Rome, but another is represented by the Villa Medicis, which for the last two hundred years has housed artists come to the city to learn more about the history of art. In the sixteenth century Montaigne remarked in his Journal that Rome was "*patched up by foreigners who have given it a lasting beauty*". In that at least, Virgil was right: great is the power of the mind over matter.

For the last few decades, man has tried to counter the destruction wrought by time by a growing awareness of what we have come to call the "*world heritage*" and a determination to protect it. But the lack of funds, political upheavals and revolutions[1] or wars[2] have already shown the limitations of these noble sentiments. While saluting the efforts of UNESCO we should observe that that organization, though it has no money, does not hesitate to increase the number of sites to be protected. Quite recently the Tassili d'Ajjer, near the Hoggar in southern Algeria, has been added to the list. There, some thirteen thousand years ago, human beings painted on walls those strange frescos we call "*roundheads*". But who will venture to protect this vast uninhabited and dangerous region[3]? Lucan may be right. These fine remains may well perish.

Indeed, human beings are most reluctant to acknowledge the marks of time and their implications.

---

[1] For example the French Revolution provoked the destruction of the magnificent Cluny III Abbey, the largest Romanesque basilica ever built.

[2] Other examples: the recent war in Afghanistan during which the giant Buddhas of Bamiyan were destroyed, or the conquest of the Inca Empire by Pisarro and his gold-crazy conquistadores.

[3] In April 2003 26 tourists disappeared and some were still missing three months later.

But not always. Take the example of Venice: the gates of the palazetti have been closed forever and the city is no longer the hive of activity it was at the time of the Serenessima. More and more frequently the Piazza San Marco is flooded, and yet ever-increasing crowds of tourists descend on the city, provoking the erosion of the buildings through the pollution they engender or the graffiti they leave behind them.

And then there are the unconscious scavengers of the heritage, like the peasant of Mongolia who builds his house with materials taken from the immense Wall of China: for him, dried mud is dried mud, an infinitely precious resource for building. Who cares about its past?

The physical matter remains while the memory fades

When we visit our old cemeteries on All Saints' Day, how sad it is to see those slabs of stone supported at each corner covering an old grave invaded by brambles. Time has caused the stone to sag in the middle and has worn away all the words once etched deeply into it by an engraver long ago. Our history is replete with this desire to immortalize a work or a life by calling on the mineral kingdom. Like the inscription engraved by Darius on a rocky outcrop of Behistun, 300 metres above the floor of the valley, to remind people of his triumphs: "*You who in times to come will look upon this inscription engraved by hammer and chisel at our command, do not obliterate or destroy it*". But time has steadily rubbed away the fulsome inscriptions desired by great men to embody their claims to immortality.

The inscriptions of Darius are fading and it is too dangerous to climb up to rechisel the lettering in the face of the cliff. And yet it is a text in three languages with a very ancient history - Old Persian, Babylonian and Elamite - and, for the Assyriologist, the equivalent of the Egyptologist's Rosetta Stone – with the difference that no one was able to read any of those three languages in cuneiform script.

Stone, a material highly valued by our distant ancestors

The characteristic feature of megalithic religions is that the ideas of immortality and of the continuity of life and death are expressed through stone[4]. Whereas the homes of neolithic peasants - who erected the megaliths – were modest and not built to last, the dwelling places of the dead were built of stone and able to conquer time. "*The rock, the slab, and the block of granite spoke of infinite duration, permanence and incorruptibility, almost a form of existence independent of future time*"[5]. Alas, no! There is indeed a continuity between life and death through stone: the matter remains but its aspect deteriorates imperceptibly. Death is written in a desert of sand become a shroud.

As the Italian artist Giuseppe Penone wrote, "*For me, all the elements are fluids. Even stone is a fluid: a mountain wears away and becomes sand. It is simply a matter of time. It is the brevity of our own existence that leads us to call materials "hard" or "soft". Deep time makes nonsense of these criteria*".

[4] Mircea Eliade, Histoire des croyances et des idées religieuses, Vol. I, p. 138.
[5] Ibid. p. 128.

Matter remains but its physical aspect deteriorates

It is an idea expressed by Ronsard in his Ode to Helen[6], but for him the matter is a living woman subjected to "*the irreparable outrage of the passing years*". Here it is the mineral world that swallows up the memory of the works of man, and his remains, in a desert of sand that becomes a shroud. The physical aspect of matter changes, but so slowly that human beings are misled. On seeing the Great Pyramid for the first time, Chateaubriand wrote: "*It is not from a sense of his nothingness that man has built himself such a tomb; it is out of the instinct of his immortality; this tomb is not the marker that signals the end of a man's brief existence but the emblem that betokens entry into an endless life*". The author of *Mémoires d'Outre-Tombe* (Memoirs from Beyond the Tomb) did not visit the pyramid itself: the Nile was in flood at the time and, not wishing to get his feet wet, the Viscount remained is his boat. His flight of fancy was most eloquent – but quite untrue: the tomb, the last of the Seven Wonders of the World, is also in the grip of time: the stones are being insidiously eaten away by the winds of the desert and by the daily swings in temperature. Sometimes, at daybreak, they sing, but in the heat of the day the edges dilate and cause a daily loss of grains in minute landslides. This relentless wearing has been calculated as amounting to 2 cm per century[7], or a metre since the time of Cheops. In 500 000 years nothing will remain. Six Great Pyramids could have been built successively and reduced to dust since the time of Lucy, our australopithecine ancestor. Further up the Nile certain necropolises of the Old Kingdom, only some 2 500 years old, are already buried under several metres of sand and, in the oldest of the pyramids, that of Saqqarah, the deterioration is much worse.

The Viscount of Chateaubriand survived without harm several changes of political regime, but knew next to nothing about the great cycle of nature: no stone can resist the alternation of desert temperatures and wind erosion. It will eventually become sand. The wind will complete this work of time and when the shroud of sand has become thick enough, the stone will be reborn in the form of a sedimentary rock.

Our intention in this book is to study the work of time, its gradual impact and the vulnerability of certain great constructions built by our ancestors. The earth has preserved only a tiny proportion of the marks made upon it by man; those made by the first ploughs, by the first horseshoes or by the wheels of ancient chariots have of course long disappeared; gone too are the scars left by violent wars, with their undermining tunnels and their trenches. How has our earth treated – preserved or rejected – the burden placed upon it over the ages by one civilization after another? Has the burden become lighter? Has Atlas wearied? Is he shirking his duty? And, if so, is he acting quickly or slowly? Have human beings become resigned to the fact? If not, what solutions have been adopted?

---

[6] We find this idea in Ronsard's sonnet, Ode to Helena. Here is an English version:

When you are old, my lady, and every evening
By candlelight do broider by the fire,
You will muse in wonder, my verses murmuring,
Ronsard sang of me when I was young and fair.
Then, no servant hearing such a strain
Tho' half asleep from toil of heavy days
Would not start awake on hearing my name
And bless your name with undying praise.
I will be under the earth, a boneless ghost
Amid the fabled shades to take my rest
You by the hearth, bent with years and pain,
Regretting my love and your haughty disdain.
Mark my words: live out your life, do not stay
Your hand, but pluck the roses of your life today.

[7] According to K.O. Emery. The wear was calculated from the volume of debris at the base of the pyramid.

Figure 1 ▶

The backing stones (the stones that lay behind the final outer casing, now missing) of the Great Pyramid[8] have failed to withstand the erosion of time.
(Photograph by the author.)

Figure 2 ▶

Undermining and deterioration of a corner of the most ancient pyramid, that of Saqqarah.

Figure 3 ▶

The religious call of the desert. At the age of 20, Antony renounced the world and set out on foot towards the heart of the desert. Later, Athanasius, Bishop of Alexandria, recounted his exemplary life. After the Edict of Constantine in 313 AD, many parts of the Egyptian desert were populated by hermits. As Claude Tillier (1801-44) wrote, "What we see as the vegetal skin of the terrestrial globe are the superposed shrouds of thousands and thousands of generations".
(Photograph Gérard Sioen, Rapho.)

[8] The facing stones were removed long after the prohibition of pagan rites by the Roman Emperor Theodosius in 392 AD. The Seigneur d'Anglure, one of the very first Egyptologists, was an eye-witness of the vandalism in around 1395, when the fine limestone facing stones were prized off and hurled down the slope. See Bonnardot, F. and Longnon, A., Le saint voyage de Jérusalem du seigneur d'Anglure, Société des Anciens Textes Français, 1878.

Sudden death

Occasionally a civilization has disappeared without warning. Many writers in the distant past were fascinated by such events: the drowning of Helike, for example, was related by Heraclides Pontius, Strabo, Pausanias, Aelian, Diodorus, Ephorus and also by Ovid, Pliny and Seneca.

Helike was an ancient Mycenaean city celebrated by Homer. It stretched inland for some two kilometres from the seashore in a plain lying between two small rivers, Selinos and Kerkynites, which flowed into the Gulf of Corinth. The city had a powerful fortress built of cyclopean blocks and nearby a sacred grove surrounding a statue of the god, master of the soil and stones, called by the Greeks Poseidon.

In the year 373 BC, as Pausanias tells us (Book VII, 24, 7), the sun shone with a reddish glow and bolts of lightening streaked across the sky. For five days, mice, snakes, rabbits and all animals dwelling underground left their burrows and fled, to the astonishment of the local inhabitants (Aelian, *On the Nature of Animals*, XI, 19). Then, one night, with an immense rumbling, the city disappeared into the sea; a huge wave swept over the land and ten warships from Sparta, anchored off the coast, were smashed against each other. Later, according to Eratosthenes, ferrymen claimed to have found, standing upright beneath the sea, a bronze statue of a human figure, which created a danger for fishermen. A reappearance of the sacred grove and statue of Poseidon. Unhappy people of Helike, victims of the anger of the god of the soil and stones.

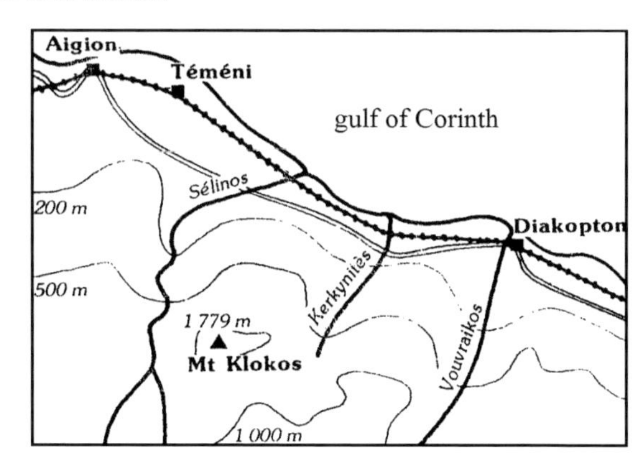

Figure 4
Site of Helike

More recent cases

Perhaps it is because they harbour the ashes of our most ancient forebears that deserts exert such a strong religious call: their immensity and their silence disrupt our sense of time so that we think of the emergence of humanoids, which in reality happened millions of years ago, as being about as ancient as three much more recent civilizations. From the time when the Sahara was a sea to the desert we know today it is no more than a few seconds for an astrophysicist, a few minutes for a palaeontologist, but an eternity for an archaeologist. For example, three civilizations of the distant past, though widely spaced in time, now seem to us to be almost contemporaneous:

The people of the Hoggar, of Tassili N'Ajjer, the so-called "*roundheads*", were a civilization that lasted some 6 000 years and covered some 350 000 km^2 with their mysterious signs; petrified settlements of 13 000 years ago swallowed up by the sand, by the oceans of dunes, and an extraordinary rock art have been found. And yet the discovery in the region of a fossilized coelacanth reminds us that the area was once covered by a sea.

In the Indus valley once existed an ancient civilization of great brilliance, judging from some of the remains of the city of Mohenjo-daro. This city was one of the 150 ancient sites scattered along the valley up to the foothills of the Himalayas. It is the one furthest downstream but even so it is over 300 km from the mouth of the river. The site reveals the superposition of three cities separated by layers of alluvial deposits, which suggest that each city was destroyed by the waters of the Indus in flood; obstinately the inhabitants had rebuilt their city on the remains of the previous one. They were stubborn but the architecture also suggests discouragement. The lowest - and so oldest - of the three cities had high thick walls enclosing a broad area, a sophisticated drainage system and even public baths, whereas the most recent one contained only makeshift dwellings. Many questions remain unanswered, which is hardly surprising since the further one reaches back in time the more mysterious are the traces. Why was Mohenjo-daro situated some 300 km from the sea despite the fact that bas-reliefs indicate the existence of trade across the sea with Sumer? Why have no ruins of coastal ports been discovered? The whole civilization faded away some 5 000 years ago, not because of the decadence that can bring down even the most brilliant culture but because of the weariness of battling vainly against repeated submersion by the Indus, which eventually buried the city under layers of sand.

Mohenjo-daro is said to compete with Ur and Eridu for the title of "*cradle of humanity*". But it is a meaningless title. We are told by the Old Testament that, towards 1800 BC, Abraham, the father of the three monotheistic religions, set out from Ur along the fertile crescent and made his way westwards towards the coast. However, two Dominicans, Father Lagrange and Father Tarragon, successive heads of the Biblical School in Jerusalem, and more recently Israël Frankelstein, director of the University of Tel Aviv's Institute of Archaeology, wishing to check the historical veracity of the biblical record, have clearly shown that there remains not a shred of material evidence for the existence of Abraham. This is unlikely to harm the three religions; it will simply be maintained that, like Antony, Abraham sought the quiet of the desert and that his ashes repose in the sands of the Syrian Desert near the Fertile Crescent.

Subjected to the erosion of time, some of the most celebrated cities of the ancient world, such as the Babylon of Nabuchodonosor II in the sixth century BC, have completely disappeared in the space of only two and a half thousand years. They have become tells, small sterile mounds with nothing to suggest the glory of their past. The remains have been reduced to a heap of rubble. If one day we inhabitants of the earth are given the opportunity to explore another planet, would our archaeologists include a person as brilliant as the German Koldewey? In 1902 he stumbled on the site of the city and with uncanny prescience selected the spots to begin the excavations that would enable him to reconstruct the most famous district of Babylon, including the Ishtar Gate, the Procession Way branching into the Babel Road beside which stood the famous Tower of Etemenanki, called by the deported Hebrews the Tower of Arrogance. More extraordinary still, Koldewey, finding the foundations of the legendary Tower of Babel still deeply scored in the earth (Figure 5) had the pleasure of noticing that his own assessments of its size agreed with the

Figure 5 ▶
Irrefutable proof of the distant past: aerial photograph of the deep scars left by the legendary Tower of Babel. (Photograph Georg Gerster, Rapho.)

Figure 6 ▶
Reconstitution of Babylon in the sixth century BC. Model on display at the Berlin City Museum. The ziggurat of Etemenanki (Tower of Babel) can be seen on the left. In the foreground is the bridge over the Euphrates.

testimony of Herodotus, who had seen the Tower itself many centuries before.

The same could be said of Amarna in Egypt, site of the short-lived capital of Akhenaton, built to escape the influence of the priests of Amon. Without the stubborn persistence of German archaeologists, it would still be buried under the sand.

Elsewhere the traces of human activity are superimposed in layers: at Hissarlik, it took Schliemann and his successors several decades to uncover the many cities, including Troy, built one on top of the other over a period of some two thousand years.

Erosion in action

Erosion by time is not always total. For instance, the lucky visitor may see emerging from the vast stretches of sand at Baalbek (Figure 7), the ruins of a rich past. "*Architecture is what produces handsome ruins*", the late Auguste Perret was fond of saying. But what becomes of those ruins? A television programme on archaeology was called "*Living stones*", but a more appropriate title would be "*Surviving stones*" since most vestiges of the past have been irreversibly mutilated.

Even taking infinite pains will never bring the past back to life. In most cases an attempt is made to focus the tourist's attention by exploiting a process called anastylosis[9]. One example is Palmyra, where innumerable sculptures lie in the sand. Some have been woken from a sleep of two thousand years thanks to the research of Japanese archaeologists. In Palmyra, the Queen Zenobia made Rome tremble and freed Egypt from its yoke before surrendering to the Emperor Aurelian in 272 AD. The city was decorated with 8000 statues, over a thousand of them in the main street. The capture of Zenobia had two consequences: she attended Aurelian's triumph shackled with chains of gold and hobbling on foot in front of the Emperor's sumptuous chariot, and had to watch him as he ordered all the columns and statues of Palmyra to be removed from sight. Poor Zenobia, admired

◀ *Figure 7*
Temple of Bacchus and Temple of Jupiter at Baalbek in Syria. Watercolour by Achille Joyau sent in 1865 to the Paris Ecole des Beaux-Arts. The stones and some of the sculptured decoration remain to remind us of the splendour of ancient temples.

[9] Reconstruction of a ruined building using for the most part materials found on the site and in accordance with architectural principles prevailing at the time it was built. The success of the operation depends on a thorough understanding of those principles.

▲ *Figure 8*
First attempt to reconstitute Palmyra. (Photograph Marc Deville, Gamma.)

for her courage in resisting a tyrannical Rome that was no longer the glory it used to be, died there in prison and was never given the opportunity to revisit Palmyra: her ashes were never mingled with the thick shroud of sand that covered a magnificent civilization; it is only recently that a forest of columns has been erected along the length of the main street.

Sometimes, but not often, the shroud of sand is momentarily stripped away by violent winds. This is what happened at Taklamakan , in the westernmost part of China at the extreme south of the steppes. The violence of the wind shifts the dunes, sometimes offering a glimpse of the treasures of small Buddhist sanctuaries.

The ruins of Hattusha (sixteenth to thirteenth century BC), once the capital of the Hittites, against whom Ramesses II fought at Kadesh, give one the impression that they are about to sink beneath the sand and are appealing for help to an indifferent humanity.

Chan Chan, capital of the Great Chimú Empire in what today we know as Peru, was a large metropolis but was gradually destroyed by sea winds, earthquakes, violent rains and looters in search of treasure. There is now nothing left to recall its bygone splendour except another field of ruins.

The highland ruins of Machu Picchu, though more recent (fourteenth to fifteenth century) are already covered in grass. As Victor Hugo wrote after the tragic drowning of his daughter

Figure 9 ▲
Ruins of Machu Picchu, Peru. (Photograph Ed. Freeman, Getty Images.)

Leopoldine and her husband at Villequier: "*Grass must grow and children must die*". Dead here too, with the last Inca ruler, are the admired remains of an astonishing civilization, which are slowly sinking into the sand. We shall see, like Marcel Proust, "*the thick grass of inspiring works on which generations of visitors will tread gaily, oblivious of those sleeping beneath, to set out their picnics*"[10].

To work our way back through the countless generations to those builders is impossible. We are confronted with the mystery of their singular lives, customs and beliefs and with the mystery of stone, "*fruits of the invisible tree of time*"[11].

My purpose in this book is to show that man's works have not always been beneficial; often the mineral kingdom has been treated as a slave, obliged to submit to the whims of man, however unreasonable they might be. Stone thus became an instrument of man's dreams of grandeur, of his idle fancies or of his gestures of defiance that were fatally flawed by errors in conception. In the beginning all was harmony and intelligence; but later an excessive audacity led to disaster.

[10] Marcel Proust, A la recherche du temps perdu, La Pléiade, Vol. III, p. 1038.
[11] Roger Caillois.

1

Do stones possess a soul? Ancient beliefs from Democritus to Lavoisier

'Be faithful to the manners of stone' (Roger Caillois)

Roger Caillois

In the exploration of my theme I have often found myself accompanied by Roger Caillois[12]. He was a man curious about everything, who applied his imagination and poetic vision to the mineral world, seeking parallels between social facts and the mineral kingdom. Stones played a large part in this curiosity, for he wrote no less than three books devoted to them "*La Pierre-écriture*", "*Pierres et autres textes*" and "*Pierres poésie*", all now sadly out of print. In his numerous studies he was generally interested in the connections he discerned between the various elements of the universe and human activities in the deeper sense (*Le mythe et l'homme* (1938), *L'Homme et le sacré* (1939), *Au coeur du fantastique*).

After reading much of this fascinating material, I wanted to continue along the path traced by this original and unclassifiable scholar using the notes I had accumulated during a long career in science followed by a second shorter career in archaeology. I wanted to show how stone had sometimes become the slave of dreams, of unrestrained ambition.

Stone in the mineral kingdom

First of all, a few words about the mineralogy, sexology and acoustics of stone will facilitate the reader's understanding.

"*O inanimate objects, do you possess a soul?*" This question by a poet was always running through the minds of our distant ancestors. They were thinking of stones which they shaped or engraved to bear, as a friend, man's first messages of hope.

---

[12] Roger Caillois (1913-1978), elected to the French Academy in 1971. International gatherings have kept his work alive: the first was held at UNESCO in May 1991 and presided by Octavio Paz, winner of the Nobel Prize for literature.

After lengthy meditation Theophrastus was drawn towards the marvellous in his work *On stones*, Plutarch wrote his *Treatise on rivers* and Pliny the Elder dealt with the subject in Book XXXVI of his *Natural History*. Nearer our time, in the sixteenth century, on the orders of the Chinese Emperor Chia-ching, the scholar Li Shih-chen produced a compilation of some eight hundred works on natural history known as *Pen-ts'ao Kang-mu* (Great Pharmacopoeia), which includes references to the fabulous stones of China and Japan.

All these ancient authors divided stones into male and female: the hardest red granites were male while the more malleable limestones were female; "*Diphye*" had the reputation of being hermaphroditic. The most masculine stones, such as the "*magnetites*", had the power to exert their will and attract all that lay before them.

Pliny assures us that stones possess a voice capable of responding to a human being; some stones were made to sing by the rays of the sun.

In the Orphic poem *Lithica*, mention is made of a stone given by Phoebus to Helenus which is treated as if it were a young child: it is dressed, washed and cradled until it lets one hear the sound of its voice.

In about 1100 AD Mi Fei, a great Chinese calligrapher, connoisseur and lover of painting, expressed boundless admiration for unusual stones and one day he put on his ceremonial robes to salute a particular stone in his garden, bowing before it and addressing it as "*Elder brother*".

Such tales remind us of the Maya, ecologists long before the term was invented, who venerated their mountains of stone. Before opening a quarry for the building of their pyramid-temples, which they knew would wound the landscape, they would offer prayers "*O Huitz-Hoh, the holy one, lord of the hills and the valleys, be patient with me... I am about to cause you harm*".

For a very long time little was known about their nature: kings, emperors and pharaohs built their tombs of stone as an expression of eternity and mystery. Goethe reassuringly called them his "*dumb masters*". Until the French Revolution nothing was known about their internal structure – it was still thought that, when stones broke, they did so without any intermediate phase, shattering with brutal suddenness. Scientists took an interest only in that sudden end[13].

It was not until the beginning of the nineteenth century that the English savant Thomas Young, able to carry out experiments with a very powerful press, revealed the existence of an intermediate elastic domain in which the cells of the stone, just before breaking, tighten up, especially in the case of softer stones; conversely, the most masculine stones tend to shatter with a loud cracking sound after very limited deformation.

So many errors have been made by architects because of their use of hard and soft stones side by side. Perhaps Roger Caillois would have drawn a parallel with social inequalities.

It is during this period of transition, which varies considerably, that the stone gathers its forces to fight against destruction. Does it suffer? Without falling for the sentimentality, dreams or veneration of ancient authors or adopting Mi Fei's ceremonious concern for his "*elder brothers*", I have always been haunted by this question.

---

[13] "*They were always assumed to be inextensible and incompressible*", in J. Peacock, Life of Thomas Young, p.421.

To set my mind at rest, I had delivered to my laboratory from Aswan a sample of the stone used for those long beams of pink granite which, nearly fifty centuries ago, the pharaoh Cheops had chosen for the five-tiered ceiling of his burial chamber. It is a stone of exceptional hardness. Aswan, where it was quarried, is located at the point where the Nile traverses its last obstacle before reaching the plain. As the cataracts are numbered from north to south, against the flow of the river, the Aswan gorge, where the bedrock is granite, is called the First Cataract. Of it Pliny says "*it is a matter to condense the entrails of the earth and to tame the assaults of rivers by placing in their path the hardest material of which the earth is made*".

In my laboratory I attached a stethoscope to this beam and subjected it to flexion. Increasing the pressure very slowly I could hear the micro-sounds, murmurs and groans preceding the powerful signals prior to the final explosion. Rather like in the Sixth Symphony of Mahler, with its tragic mood and gradually mounting crescendo, there can be no other issue for the hero but death. Microphones with amplifiers confirmed what was happening but I always came back to the stethoscope, that brilliant invention of Dr Laennec, which can locate the source of warning sounds within the patient's chest and which gave my research a human touch: there was no doubt that stones express their feelings, and their simple nature, on which human beings so often rely to perpetuate their own deeds, is continually exploited by man.

Then I recalled the impassioned arguments of the celebrated Dominican, Bartolomé de Las Casas, defender of the Indians being massacred by Spanish and Portuguese conquistadores during their conquest of the New World. To cleanse themselves of their turpitude they claimed that it was very hard to distinguish body and soul among the Indians, even assuming that they possessed a soul: they were regarded as "*higher animals*", as creatures resembling human beings, as denizens of the devil's empire; it was therefore permissible to torture them unscrupulously in order to ensure that they enjoyed a better life in the next world.

Stones are in much the same situation. They are of two kinds: first there are the humble rocks that one can find anywhere and use to construct modest buildings or drystone walls. Such walls have always fascinated craftsmen throughout history: I remember reading as a child a book that told the story of a young mason who, before laying the next course of stones for the wall he was building, waited until the stones in place had told him what they wanted.

The second kind are considered more noble, are often brought from far away and are renowned for their quality; skilful stonecutters make them suffer but with the praiseworthy design to bring out their beauty. They cut them into mighty blocks, called by the Romans *lapides quadrati*, and then shaped and chiselled them; the stones were thus assured of a fruitful life, almost eternal, moving on from a castle left in ruins after a siege or from a church that had collapsed to a palace that was subsequently burnt down and then to the town hall of a young republic. The stones recovered after the demolition of the Bastille in Paris were used by the engineer Jean Rodolphe Perronet to construct the Pont de la Concorde during the French Revolution "*so that the people may continually tread underfoot the stones of the ancient fortress*". That great servant of the state, the first *Grand Ingénieur du Roi*, who had served the king for over fifty years[14],

[14] In 1750, Louis XV created a 'roads and bridges' service and the still-existing Ecole des Ponts et Chausées, which Perronet directed until his death.

found the right words and succeeded in entering the good graces of the Revolutionary Committees, thus avoiding the fate of the chemist Lavoisier, to whom the executioner coldly observed "*The Republic has no need of savants*".

In 460-370 BC, in ancient Greece, lived the philosopher Democritus. As Claude Allegre underlines in his recent book, "*he had understood or at least glimpsed the meaning of everything*". For him, nature is composed of atoms, individual particles that are material, indivisible and invariable; entities are born out their combination and whirling motion. After him, there were centuries of contradictions and his ideas were forgotten. For defending those ideas Giordano Bruno was burned at the stake in 1600 and Antoine Lavoisier, the father of modern chemistry, found no one to protect him from the guillotine in 1794. He had been resolutely on the side of Democritus and had just pronounced his famous dictum "*nothing is created and nothing is lost*" to the glory of atoms.

People pass by admiring the final destiny of these immortal stones. Those used for the Pyramid of the Niches at El Tajin[15] in Mexico, for example, were, in the words of Octavio Paz, animated by a breath of life: "*Its stones are alive and dance; it is neither petrified movement nor suspended time: it is a dancing geometry*". And this dance has been continuing for over a thousand years.

All these stones mean more to me than the emeralds, rubies, diamonds and other "precious" stones worked by jewellers, which are often impure in that they contain micro-inclusions in the form of crystals, gases or organic particles that dull their brilliance.

But let us turn our attention to those other beautiful stones that were the slaves of man's dreams of grandeur. Their history is often sad and sometimes tragic. My wish is to share these feelings with the reader, so that he or she will be able:

- To imagine the explosion, with a noise greater than the firing of a cannon, that accompanied the sudden rupture of the quintuple roof of stone beams above the King's Chamber in the Great Pyramid of Cheops towards the end of his reign;

- To share the terror of an eighteenth dynasty pharaoh's son who was bound to the tip of the obelisk being erected to the glory of the king his father;

- To feel the disappointment of the proud Ramesses II lying in his tomb caught up in a *danse macabre* because it had become a cesspit for the valley of the kings;

- To hear the cries of the faithful in St Peter's Cathedral in Beauvais on 18 April 1572 when, at the moment the priests were making themselves ready to issue forth in procession, a master-builder yelled a warning to everyone to save themselves; hardly had the clergy and parishioners run from the church when the spire, which was higher than the Great Pyramid, and a large part of the pillars and vaults came crashing down to form a great mound of stones;

---

[15] The Pyramid of the Niches at El Tajin, Veracruz, Mexico, built in the seventh century, had seven levels and 365 recesses.

- To share the terrible anxiety of those trying to save the Leaning Tower of Pisa when it was within an ace of collapsing in September 1995 – still called 'black September';

- To watch the hearse of Soufflot, dead of chagrin in 1780 after a silly wager, and to examine the *Journal de Paris* of 19 Vendemiaire in Year 5 of the Republic which enumerated the faults of the Pantheon and announced that there was no hope of the building surviving for posterity;

- To share the joy of Franck Goddio when fine statues, coins and the Stela of Nectanebo I were lifted from the waters of Aboukir Bay in Egypt.

Reader, stay by my side.

2

The song of stones

We are told by an ancient Chinese scholar named Li Shichen, that there exists a cave in the west of the prefecture of Chi called the Cave of Dragons and Fish. In it there is a stone that is sometimes big and sometimes small; if it were split in two one would find inside the shapes of dragons and fish. People passing by the cave refrain from talking; they hear the distant sound of thunder and hurricane and are stricken with terror.

Pliny stated that stones possess "*a voice that echoes the voice of man*". "*The Greeks*", he added, "*call this phenomenon an echo. At Olympus, men of great ingenuity have taken remarkable advantage of it to build the portico, which is called the heptaphone because it sends back the same sound seven times*".

In fact, the ability to respond to the human voice is possessed only by the hardest stones and depends on the geometry of the site. This explains the prodigious effect of the low vaults[16] in the Cistercian abbey of Thoronet: the hard limestone of which they were built throws back an echo of the Gregorian plainsong chanted by the monks that lasts for eight seconds and sometimes longer, as if some piece of software in the body of the stone had first recorded and then repeated the orderly and musically perfect chanting. "Listen, you will see" the founding abbot Bernard de Clairvaux appeared to be saying.

Purity of sound: a miracle of these handsome vaults inspired by the faith of monks who, as they entered the monastery, would start to chant the plainsong *Ergo sum resurrectio*, a melody filled with the hope of lifting the soul to contemplation of the sacred.

Amazed by such marvels in sound, builders dismantled the timbered roofing of the churches, for they had found that a stone barrel vault allowed the liturgy to resound much better. In this respect, Romanesque architecture kept its superiority over the subsequent Gothic. Indeed, experiments conducted by the choir of the Cathedral of Dijon have shown that the same plainsong sung under the low vaults of the Beaune Collegiate Church resounded better than under the ribbed vaults of the Cathedral of Dijon, where the voice of the stone was mingled with echoes and interference.

[16] In Cistercian practice, the haunches of the vault were only 2.90 m above the floor. Under the Rule of Saint Bernard, moreover, there was almost no sculptured decoration or furniture, circumstances which improved the quality of the sound.

Figure 10 ▶
The reddish stones of the cloister of the twelfth century Abbey of Thoronet, Var, France.

Most ancient theatres are now mere heaps of ruins but we can still, during the summer festival at Orange for example, imagine how their architecture swelled the voice of the tragic actor.

There is at least one place where reality comes to the aid of the imagination: it is the King's Chamber in the Great Pyramid in Egypt, the holy of holies of the largest ancient monument that has survived to our day over nearly forty-five centuries. When one enters, one is impressed by the harmony of the proportions[17] and the nobility of the red granite that covers the walls and ceiling – "*the finest specimen in the world of polished granite worked down*", as the nineteenth century archaeologist Howard Vyse noted. It enables sound waves to reverberate with an increase in loudness, producing a most impressive acoustic effect. Just before I entered the Chamber for the first time, there was a minor accident that fused the electrical lighting, leaving the interior of the Great Pyramid in total darkness. There was just the sound of a group of Muslims who continued their chanting in the King's Chamber. The sound reached me as I stood at the top of the Grand Gallery with wonderful purity and loudness. When the electricity came back on I walked into the Chamber and was amazed to find a single Muslim humbly prostrated in a corner. Astonishing amplification.

Graham Hancock[18] thus describes his impressions: "*I know not how long I remained in the chamber. The atmosphere was dank, the air warm, as if exhaled by some enormous creature: I worked my way to the centre of the chamber and made a low hum; the walls and the ceiling seemed to gather together and expand the sound waves and then send them back to me; I felt a tingling in my legs. I was electrified… and at the same time calm as if about to be vouchsafed a frightening and inevitable revelation*".

Even today, although its walls are soiled by the deposits from the breath of too many visitors, there is a powerful reverberation in the Chamber. At the time of Cheops, when the stone roof-beams had been newly cut and polished with a perfect finish, the energy of the reflected sound would have been much more powerful.

---

[17] The proportion of length to height is close to the golden section.
[18] *The Sign and the Seal ; a Quest for the Lost Ark of the Covenant.*

Are the stones trying to tell us about the tragedy that occurred as the construction of the Great Pyramid neared completion, when all the massive stone beams that roofed the chamber and the compartments above suddenly ruptured with a terrifying crash? There seems to be a presence in this chamber and in its stones, which might explain the sometimes odd behaviour of certain visitors. One winter evening I was present at closing time and the numerous visitors were drifting away with murmurs of regret when one of them obstinately rebelled and lay down in the sarcophagus. Wishing to spend the night there he offered the guardian a large tip, for he had been told that, if he lay in a north-south orientation in the sarcophagus, special waves like those that animate a compass would pass through and reinvigorate his body. True or false? Certain ultra-hard stones, which the ancients used to call *magnetites*, are slightly magnetic – and hence have two poles - and it is possible that the roof-beams, laid in a north-south direction pointing towards the magnetic pole had been arranged like a set of torch batteries with their poles alternating to amplify the strength of the original magnetic current and by induction pass through the sarcophagus and revitalize a person lying in it.

The singing stone

We are reminded of the two famous colossi of Memnon (Figure 11) which at daybreak used to emit a vibration, the "song of Memnon". These huge statues of stone, over fifteen metres high, had been erected in front of the temple of Amenophis III (1408-1372 BC), the father of Akhenaton and a pharaoh known for his lavish style of life. Placed facing the rising sun, the colossi have their backs to the mountain within which, in the setting sun, the pharaohs were buried for centuries. In their imposing and melancholic solemnity they now sit before a field of ruins, the remains of an outstanding temple to the glory of Amenophis III.

◀ *Figure 11*
The colossi of Memnon.

It was the Greeks who identified these statues with the hero Memnon, king of the Ethiopians, who had come to the aid of besieged Troy: according to the legend, he was the son of Zeus and the Goddess Aurora and his body was restored to life each morning under the caresses of his mother. Later, under the Romans, the reputation of the colossi was still great: tourists of the time covered them with inscriptions, as Tacitus remarks in describing the voyage of Germanicus[19]. When Hadrian visited Egypt in 130 AD accompanied by the Empress and the handsome Antinous, who would be drowned accidentally in the Nile shortly afterwards, the Emperor was taken to see the famous statues. But they remained silent and it was several mornings before they deigned to acknowledge the Emperor's presence.

I have always enjoyed this anecdote. Stones, as I have said, have been enslaved by man, who has shaped them mercilessly to his will. In this story there is an act of insubordination: the colossi snub the presence of an emperor and do just what they want!

Each of the colossi was hewn from a single block of pink quartzite. The rising sun would strike them with its rays, which warmed the stone. It was the more northern statue which used to emit the sounds, as we can see from the inscriptions left by tourists. Many recent writers on the subject, particularly R. H. Wilkinson in his book *Egypt*, refuse to credit the tale, seeing it as a "trick of the priests", who must have struck the statue from behind with a special stone. This peremptory opinion reflects a quite common attitude in the history of science and technology, which consists in completely rejecting ancient beliefs while at the same time refusing to engage in any rigorous analysis. In this particular case, research in the laboratory has helped us to understand why the colossus of Memnon emitted sounds in the morning. The stones contain water. Roger Caillois, in his book *Pierres*, speaks of a nodule of agate : "*If you shake it close to your ear, it occasionally, but very rarely, lets you hear the sound of water lapping against the walls inside*". "*Undoubtedly*" he adds, "*there is water within the nodule, imprisoned in a gaol of stone since the birth of the planet. The desire awakens to set eyes on this imprisoned water*".

Stones do indeed have pores of varying sizes[20] that contain water, but that water is not necessarily imprisoned. A homogeneous block of quartzite, like that of the "babbling" colossus, would contain pores that were mainly vertical and *perfectly unbroken*, within which water would rise by capillarity in the morning charged with crystals from the groundwater. These crystals would expand when warmed by the sun, making the stone emit a humming sound. The process has been duplicated in laboratory experiments[21].

When the colossi were damaged by an earthquake, the continuity of the pores was ruptured. In around 200 AD Septimus Severus tried to restore them but in vain - their song had ended because the shock waves of the earthquake had broken that continuity and silenced the voice of the stone for ever.

---

[19] Tacitus, *Annales*, 2, 61 ; see also A. and E. Bernard, *Les inscriptions des colosses de Memnon*, Cairo, 1960.

[20] The porousness of stones - the proportion of empty space within the total volume – varies from a few percent in the hardest stones to around twenty-five percent in softer ones.

[21] Recently, several others have been unearthed near the two colossi, but the author does not know whether or not they are able to sing.

The resonance of stone

Certain medieval "companions of the tour of France", apprentices who would become the future master-builders of churches, were able to make stones sing as they chiselled them into shape; the sound made by the travertine used to build Chartres Cathedral was unique. Stone-cutters knew how to make all fine stones sing.

Dodona, in Epirus, was famous on account of its oracle of Zeus. Its priests went barefooted because, they said, they were better able to feel the vibrations of the divine voice and thus keep in close contact with the god. They frequently twisted his message for the crowds of pilgrims who hastened thither to hear the predictions of this benevolent god. In this case it was in fact the resonance of a rock which the crafty priests used in order to exploit the credulousness of ordinary people.

Figure 12 ▲

View of the Greek theatre at Dodona in Epirus, celebrated for its oracle of Zeus. Its priests went barefoot in order to be in close contact with the god.

The siege of Appolonia

Much more serious, and with tragic consequences, were the techniques of undermining used in the most ancient strategies and which prompted responses of great ingenuity on the part of savants familiar with the physical properties of stone. One of these experts was Tryphon, the architect of Alexandria. The story is told by Vitruvius in his *De Re Architectura* (Book X, chapter XVI):

"*When also the city of Appolonia was besieged, and the enemy was in hopes, by undermining, to penetrate the fortress unperceived; the spies communicated this intelligence to the Apollonians, who were dismayed, and, through fear, knew not how to act, because they were not aware at what time nor in what precise spot, the enemy would make his appearance. Tryphon of Alexandria, who was the architect to the city, made several excavations within the walls and, digging through, advanced an arrow's flight beyond the walls. In these excavations he suspended brazen vessels. In one of them, near the place where the enemy was forming his mine, the brazen vessels began to ring from the blows of the mining tools which were working. From this he found the direction in which they were endeavouring to penetrate, and then prepared vessels of boiling water and pitch, human dung, and heated sand, for the purpose of pouring on their heads. In the night he bored a great many holes, through which he suddenly poured the mixture and destroyed those of the enemy that were engaged in this operation*".

This knowledge of the resonance of rock led the Greeks to protect their most beautiful monuments of stone against earthquakes: the temple of Artemis (Diana to the Romans) at Ephesus, rebuilt after a fire and considered by Pliny (XXXVI – 95) one of the Seven Wonders of the World, was built "*on marshy terrain so that it would not feel the shocks of earthquakes or be endangered by crevasses in the ground*". And to prevent it from slipping down the slope, a bed of compacted carbon was spread over the site and covered with fleeces to make an excellent anti-vibration carpet.

The grumbling of cliffs

I shall never forget the marine cemetery at Varengeville-sur-mer. It stretches along a steep cliff between a charming little church and the sea. Numerous artists, unaware of the implacable processes of nature and the work of time, chose it for their eternal repose. But the sea below was endlessly bombarding the chalk cliff with pebbles, wearing away its base and causing chunks of the cliff to fall into the water, so much so that some of the tombs are now hanging over the edge of the cliff. Woe is man! I was consulted as to what should be done but refused the task partly because I knew full well what little man can do against such phenomena but also because I was reminded of the Romans who, to guide their ships in the English Channel, had built watchtowers well behind the cliffs.

One of these watchtowers was the Tour d'Ordre at Boulogne-sur-mer, whose name is a corruption of *turris ardens*. Said to have been erected by Caligula, it was restored by Charlemagne but parts of it collapsed into the sea on several occasions between 1640 and 1645 (Figure 13). One of many forgotten disasters. It was originally situated about an arrow's flight from the edge of the cliff. Octagonal in shape, with a circumference of 192 feet, it had twelve floors each one set back a foot and a half from the one below. The tower became a victim of the sea, which gradually ate away the cliff, and of the insidious undermining wrought by springs and quarrying. But its collapse was due above all to the outrageous indifference of the aldermen of Boulogne.

Large segments of the chalky cliffs bordering the Channel collapse into the sea. In places the land is retreating by 30 cm a year. Nowadays, as a disaster beckons, people lend an ear to the sounds made by the cliff. The chalk is not silent: the blocks shift in the vicinity of fault-lines all over the cliff. At night, when the weather is calm, one can hear cracking sounds: the placing of geophones – an instrument that detects vibrations in rocks - at various levels will warn watchmen of an imminent danger. But what would have been the use of geophones for the illustrious dead in the cemetery of Varengeville? Fate was taking a hand well beyond their death.

Figure 13 ▲

The Tour d'Ordre at Boulogne-sur-mer before it collapsed into the sea towards 1640. Upper left: the tower besieged by the English in 1509.

3

Stone as a messenger

In all ages and on all continents man has confided his knowledge and his memories to stone. But even earlier, in the Palaeolithic (literally 'ancient stone'), *homo habilis* sought stone of the finest quality to fashion his first tools. By brutally striking one stone against the other he occasionally succeeded in producing a sharp edge. *Homo habilis* was not very skilled in this work for it has been calculated that a kilo of flintstone produced little more than ten centimetres of cutting edge. Later, *homo erectus* succeeded in producing four times as much and *homo sapiens sapiens*, about a hundred thousand years ago, seven hundred times as much. Yves Coppens, author of this curious statistic, thinks that it provides a criterion for measuring the progress of man's intelligence, which evolved in exponential fashion. An intelligence that kept in step with his cruel destructiveness of stone. Prehistoric man, not a very aggressive being but wishing to express himself, used the cutting edge of one stone to engrave another. In other words, art and writing were born out of this fratricidal confrontation between stones.

◀ Figure 14

Toolbox of Upper Palaeolithic man: flints for various purposes.

Prehistoric man used caves as places of refuge, and it is mainly on their walls that he expressed his art. But nobody is able to decipher the exact meaning of all the messages transmitted by this rock art. We find a mixture of beliefs, preoccupations, imagination and religion in the 700 000 identified sites scattered over all five continents and from which a total of around 30 million images and signs executed during the last sixty thousand years or so have been collected. In Europe, over two hundred caves are decorated with rock art, 85% of them in southern France and northern Spain.

One motif encountered in all decorated prehistoric sites is the outline of a hand. In another context, Herodotus[22] tells us that the hand on a stone authenticated an oath: "*No nation regards the sanctity of a pledge more seriously than the Arabs. When two men wish to make a solemn compact, they get the service of a third, who stands between them and with a sharp stone cuts the palms of their hands near the base of the thumb; then he takes a little tuft of wool from their clothes, dips it in the blood and smears the blood on seven stones which lie between them*". The stone which heard an oath served as a witness among the ancient peoples of Palestine and Transjordan.

In cave art, certain prehistorians discern the concept of art for art's sake, the desire for beauty, while others stress the idea of an art dedicated to fecundity (hence the "Venuses" with their bulging bodies) and thus to the preservation of the species (hence the great predators pierced with lines), while yet others, like Abbé Breuil, emphasize the magical and religious context or, like A. Leroi-Gourhan, their function as 'mythograms'.

Little by little, in the open air, artists began to use flints to engrave on small tablets of limestone representations of their fellow human beings. These images are already a little like family portraits. Much later man discovered writing, and that was really when stone acquired its letters of nobility;

Figure 15 ▶

The hand, a common motif in prehistoric cave art.

[22] Book III, 8.

it became the medium for creation and for thought or, more prosaically, the necessary tool for measuring, sharing and distributing material riches. However, there is a striking contrast between the richness of a sophisticated spiritual life and the very simplistic command of mathematics: the Egyptians, for example, never found out how to represent multiplication or division by 2.

Pictograms
Ideograms
Psychograms
Horse
Labia
Rectangle wiltrays

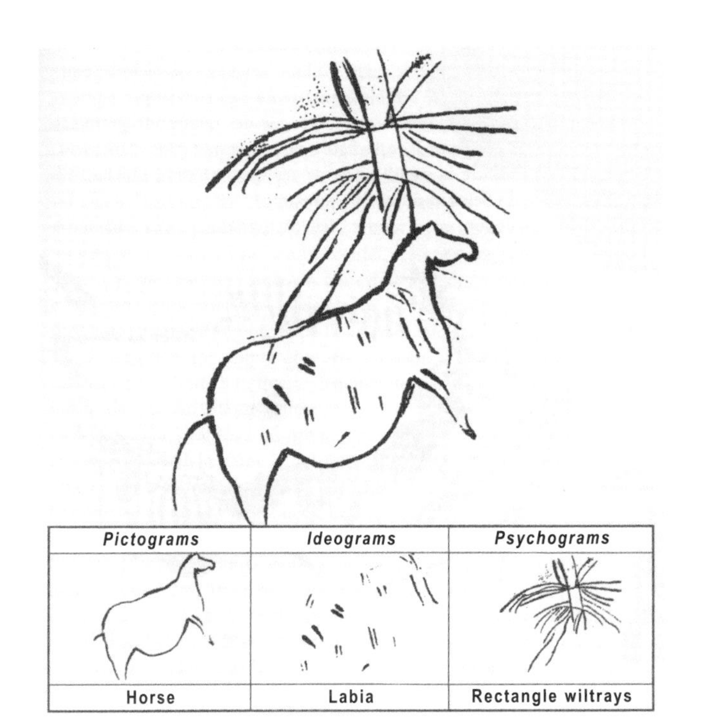

◀ *Figure 16*

The message of the horses of La Pileta (Spain): the three major categories of signs are pictograms, ideograms and psychograms[23].

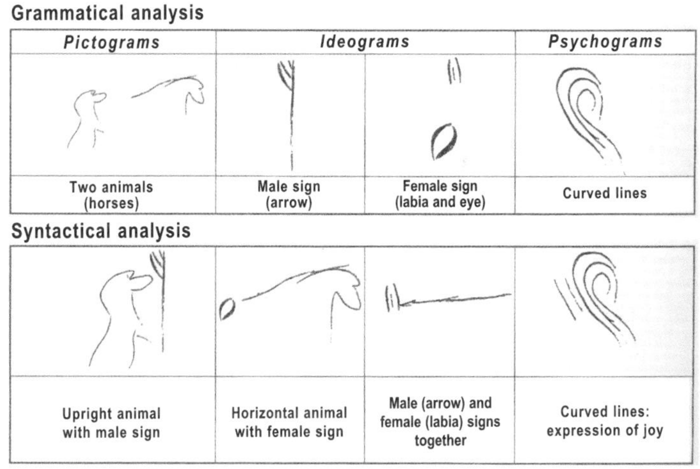

Grammatical analysis
Pictograms
Ideograms
Psychograms
Two animals (horses)
Male sign (arrow)
Female sign (labia and eye)
Curved lines
Syntactical analysis
Upright animal with male sign
Horizontal animal with female sign
Male (arrow) and female (labia) signs together
Curved lines: expression of joy

◀ *Figure 17*

Other examples.

[23] Pictograms are figures that resemble identifiable animals, humans or objects; ideograms are repetitive signs of a synthetic nature - there are about twenty of them throughout the world; psychograms are signs functioning at the subconscious level and expressing the personal feelings of their author.

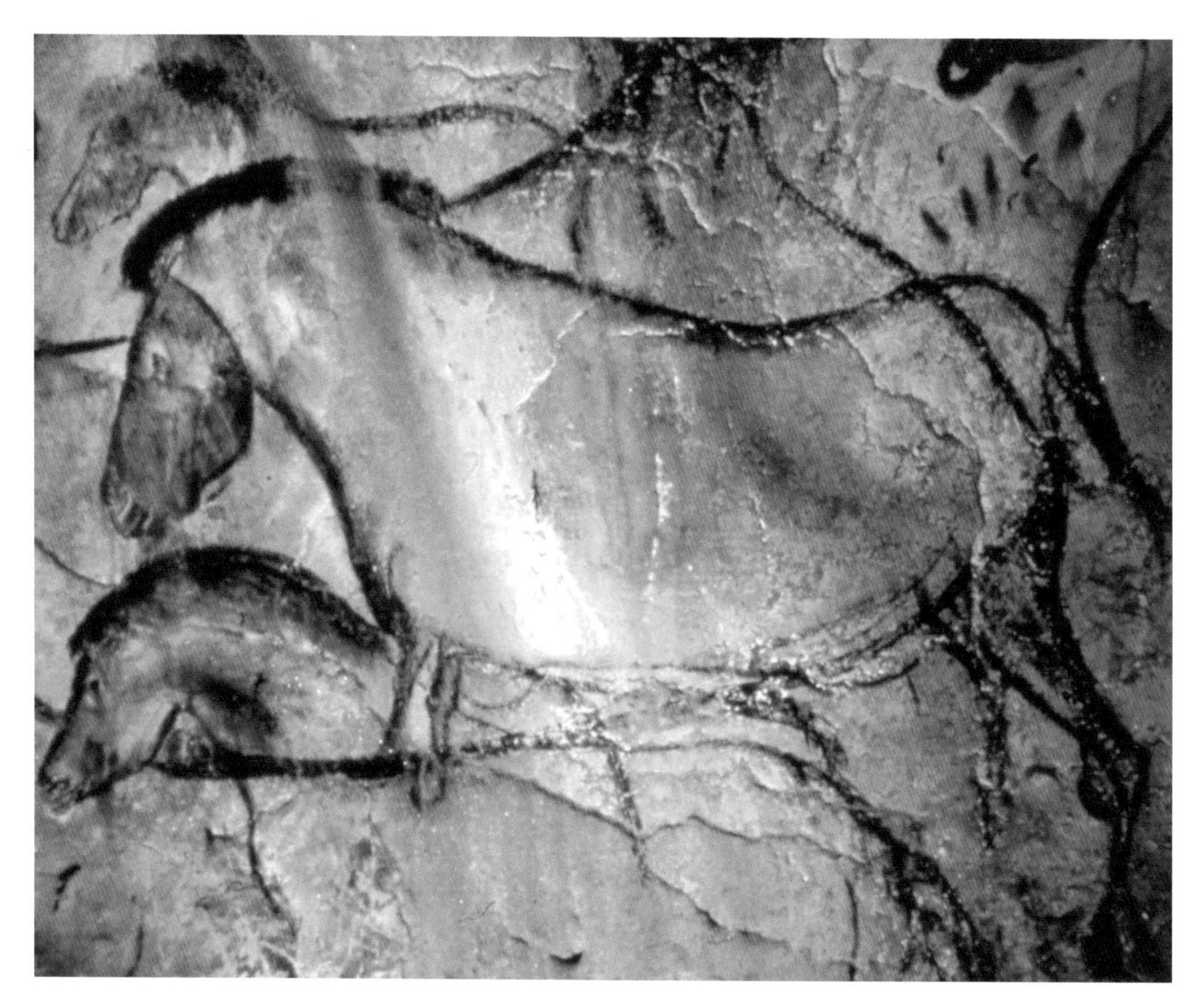

▲ *Figure 18*

Silhouette of a horse in the Chauvet cave (Ardèche, France), traced some 30 000 year ago. The Cro-Magnon were humanity's first great artists. (Photograph French Ministry of Culture.)

In Egypt, as we shall see, the obelisk was a form of writing tablet which served to give an account of the life of the pharaoh, with the inscriptions on the four faces of the monument pointed at the sky recounting his exploits and exalting his glory. Certain kings were extremely vain and went so far as to cover the four sides of their monuments with what we call hieroglyphs. As Champollion would write in 1824 in his *Précis du système hiéroglyphique des anciens égyptiens* (Primer of the hieroglyphic system of the ancient Egyptians): "*The royal glyphs [are] at once figurative, symbolic and phonetic in the same text, in the same phrase, in the same word*".

In the future tomb of the pharaoh, artists had to create a veritable fresco on the walls before his death. The tomb of Ramesses VI (Figure 19), for example, contains the most complex wall painting so far known.

Figure 19 ▲

Decorated wall in the sarcophagus chamber of Ramesses VI, a long summary in hieroglyphs of all the beliefs and myths; it seeks to explain the great mysteries of death and rebirth.

In Egypt, as in Mesopotamia, the writing on the stone was done by scribes, who perfected the techniques, adapted and disseminated them as necessary, and passed them on to the next generation.

In one Mesopotamian document, a scribe talks proudly of his art to a colleague: "*I know how to write on tablets, on tablets that measure from one to six hundred gur of oats, on tablets that give weights from one sicle to twenty mines of silver, I know how to write the marriage contracts or business contracts I am entrusted with,... contracts for the sale of land, of slaves, pledges in silver, tenant farming contracts, contracts for the growing of palm-trees, and even adoption agreements. I know how to write all that*". As Mesopotamia had no stone, the scribes had to use clay tablets: their signs were inscribed with a sharpened calamus reed on the soft surface of the clay before it was fired.

The Egyptian scribes remained loyal to hard stone that preserves for ever the imprint of the sign or image they chiselled. Learning to write was a long and difficult process, but in a country in which the mass of the population was illiterate, it was a skill that gave access to a high social standing: in Egypt, it took about ten years to learn the numerous hieroglyphic signs. In order to train their memories and learn to write, the scribes would chant together; they also used textbooks, collections of model letters and dictionaries that classified words by subject. Their schools were attached to the Palace and to the '*Houses of Life*' of the temples, which also served as libraries, archives and a kind of university. The scribes were the pillars of intellectual life and future copyists of the sacred texts. Numerous texts in praise of the profession have been found in Egypt. Here is one: "*It protects from toil and from all labours; it dispenses you from carrying the hoe or the pickaxe; you do not have to lug baskets around; you are exempt from pulling oars. You are spared anguish: you are not under the orders of numerous masters, of a multitude of superiors. For of all those who exercise a profession the scribe is the most important*".

A cushy job, it would seem, for the scribes; but those who succeeded were true sculptors and masters of the techniques of colour. How could we fail to be touched some twenty years ago by the

Figure 20 ▶
Egyptian scribes doing the accounts of a funerary domain. Mastaba of Akhethotep, 5th dynasty of the Old Kingdom (2450-2290 BC).
(Photograph Daniel Lebée, RMN.)

◀ *Figure 21*
The celebrated scribe Ahmes, with the piercing eyes (ca. 2500 BC). In the white limestone of the cornea is inlaid a cone of hollowed out crystal with the point directed inwards: the Egyptians were familiar with the anatomy of the eye and reproduced faithfully the curvature of the cornea and the diameter of the pupil. In the Louvre in midsummer the eyes of Ahmes glitter in the rays of the setting sun.
(Photograph Franck Raux, RMN.)

beauty of the texts in the burial chamber of the small pyramid of Unas? The inlaid colours, particularly the blues, were magnificent and each hieroglyph a work of art. But most of that brilliant colour has lost its gleam: the arrival of innumerable tourists in such a tiny space has sullied the walls, a bit like in Fellini's film *Roma*, where the frescos painted on the walls of the underground Etruscan hall, found when tunnelling for the Rome metro, rapidly faded as a result of the pollution introduced by too many visitors.

In Egypt the hieroglyph, which was reserved for a privileged class, was later replaced by other symbols easier to decipher, a demotic script written on papyrus made of reeds, but stone long remained a medium of communication, as is shown by a recently discovered Coptic ostracon.

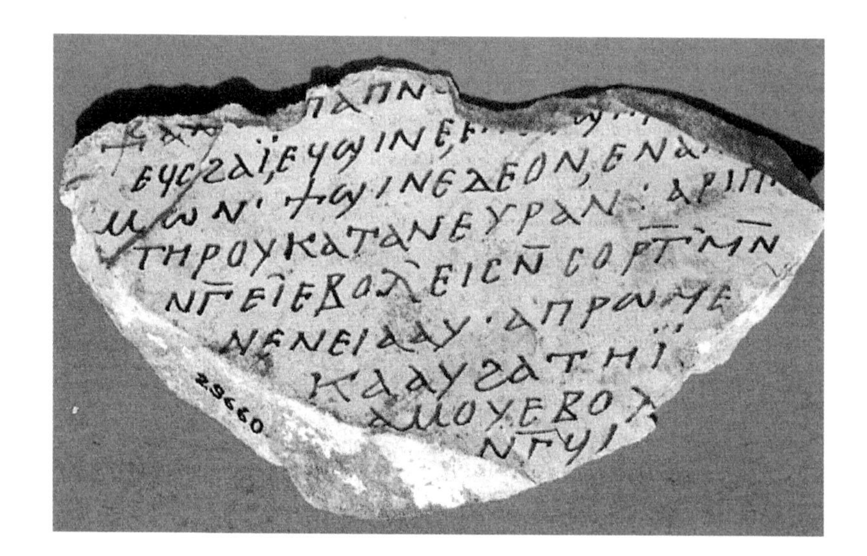

Figure 22 ▶

Coptic ostracon discovered recently. Note how the writing system has been simplified.

Figure 23 ▶

An epic engraved on a stone (1m80x1m30) relating the life of a Mayan King.

Among the Maya certain epics were engraved on stone. At Piedras Negras, in the north-east of Guatemala, archaeologists have recently brought to light a sculptured and engraved stone panel measuring 1 metre 80 by 1 metre 30, a stela that bears a long story recapitulating the life of a Mayan king who died not more than 1 300 years ago. He was called *Itzamk'anahk K'in Ajaw'* and reigned for forty-seven years. The uniqueness of the text lies in its exceptional length: it is five or six times as long as most other texts found hitherto. There have always been great chatterboxes wanting to talk about themselves on stone.

◀ *Figure 24*
The Bibliotheca Alexandrina seen from the south looking northwards. "Like the solar disk surging out of the night to illuminate the world of knowledge" said the enthusiastic Mohsen Zahra[25]*, director of the project designed by Norwegian architects.*

◀ *Figure 25*
The windowless south wall of the Bibliotheca Alexandrina on which examples of all the world's writing systems have been engraved on Aswan granite. (Photograph Stéphane Compoint.)

Also to be found among the Maya, at Palenque, is the famous Pyramid of Inscriptions on which 620 glyphs have been engraved.

Throughout history persons imprisoned in stone cells have confided their unhappiness and misfortune to the walls around them. "*Les murs étouffent les sanglots et éteignent l'agonie*" (the walls smother our sobs and extinguish our dying breath). These words were engraved on the wall of the secret prison hidden within the thickness of a wall in the tower of Loches, near Tours in France[24]. Consolation in stone!

No greater tribute could be paid to stone as the bearer of messages than the one imagined by the creators of the new *Bibliotheca Alexandrina* built in memory of and to replace the one built by Ptolemy I in the third century BC, which contained some 700 000 papyri. The roof of the new

---

[24] Museum dedicated to the memory of stones at Verneuil-en-Halatte in the Oise département.
[25] More mischievous spirits see it as a reminder of the enormous tsunami that, on 21 July 365 AD, struck the city. Five thousand inhabitants lost their lives. The tragedy was commemorated for two hundred years. The giant wave swept over the tops of the houses bordering the sea and its backwash caused them to tilt towards the sea.

library is sloped northwards towards the sea. The south wall is windowless, 32 metres high at its highest point, and covered with hundreds of small slabs of granite from Aswan 15 cm thick on which 6 300 forms of writing or signs have been deeply incised to represent all periods and all known cultures of the present-day world.

4

Stone obelisks to the glory of the all-powerful

Since the earliest times the Egyptians, whose land is chock-a-block with fine stone, took up the idea of cutting a stone in the shape of a needle and then erecting it to point towards the sky as a tribute to Re-Harakhti, the god of the rising sun. The upward movement and fine polish of these monuments aroused admiration.

The simple drawing in Figure 26, from chapter 15 of the *Book of the Dead*, a text whose purpose was to accompany travellers to the next world, is entitled "*Worship of Re-Harakhti as he rises on the eastern horizon*". The scene shows two priests, one reading a papyrus which he is holding in his hands and the other making offerings to the two obelisks representing Re-Harakhti.

As the pharaoh was identified with Ra, the sun, these needles became, as we have seen, a kind of writing tablet on which to record his life, with hieroglyphs on all four faces recounting his immortal glory.

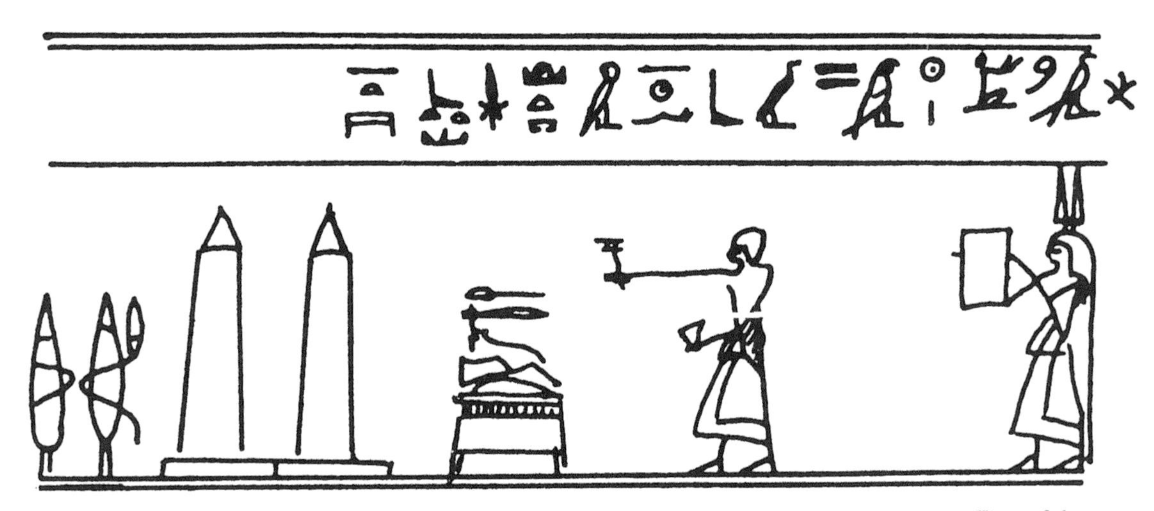

Figure 26 ▲

Sketch taken from the Book of the Dead.

Naturally, the higher and the thinner the obelisk the more fitting it would be for the exaltation of a pharaoh and his reign. But to put this idea into practice, what stones should be used? For the ancient Egyptians – and for other peoples too – stones were regarded anthropomorphically: for a pyramid a masculine stone was preferable, but which?

Certain of the stones at their disposal seemed to them to possess eternal life: these stones were of nummulitic limestone. Nummulites were tiny creatures that had lived in their billions in the warm seas of forty-five million years ago. Before the sea evaporated, their bodies had settled on the bottom, compacted and eventually turned into the limestones of the plateau of Gizeh, the base on which the three large and heavy pyramids were constructed. They were stones that had once been alive and had preserved more successfully than a mummy the memory of that first life.

It was only during the reign of the pharaoh Chephren that it was observed that these limestones were not always perfectly homogeneous, being sometimes infiltrated by veins of marl. Chephren had given orders to hew out an enormous block of nummulitic limestone in which an unknown sculptor had then carved the Sphinx, that majestic hybrid of lion and man that, we are told by the *Book of the Dead*, "*keeps watch on the edge of eternity … over everything that was and everything that will be*". Alas, the giant block was not as pure as nummulitic limestone usually is: some veins of marl had got amongst the nummulites in that warm sea, with the result that today we have the impression that the face of the Sphinx is full of scars or wrinkles in the vicinity of the lips and eyes.

The pharaoh and his sculptor had not made the right decision when they chose the block to be carved. All those who tried to raise long obelisks hewn out of this stone probably encountered the same problems. After a number of failures the choice fell exclusively on the granite of Aswan despite the fact that it was so far away. These rocks of Aswan were born in the furnace of the earth, male among males, the hardest of the hard. The pharaohs then competed in the height of their obelisks,

Figure 27 ▶
The nummulitic limestone of Gizeh enlarged fifty times.

Figure 28 ▶
The Sphinx: the veins of marl that infiltrated the limestone of the plateau seem like wrinkles in the vicinity of the lips and like scars around the nose and the eyes.

which rose to just over thirty metres in the eighteenth dynasty – more precisely 32.18 m at the end of the reign of Thutmosis III. Egypt was then at its zenith, but that particular obelisk was rudely treated: it was first of all erected at Karnak before being transported to Constantinople and finally to Rome, where it fell and broke when its pedestal gave way. It was eventually re-erected in Piazza San Giovanni Laterano in Rome.

In this race to get the highest obelisk, the kings of Egypt had to overcome a number of set-backs: the unfinished obelisk at Aswan is visible proof that lofty ambitions can fall at the first hurdle. This monolith was planned to be 41.75 m high, rising from a square base with sides of 4.20 m. It would have weighed 1 168 tonnes, a load extremely difficult to haul, to transport and then to erect! But those problems never arose because cracks appeared in several places within the rock while it was being dressed in the quarry prior to extraction. And so the pharaoh gave up despite the fact that two side trenches had already been excavated with great difficulty using balls of dolerite. As they contemplate this sleeping giant, generations of tourists give only an absent-minded glance at the two great balls of dolerite weighing some 500 kilos each with which, under a fierce sun, the fellahin spent day after day striking the stone in order to extract the obelisk.

Aswan lies at the foot of the granite massif of Ethiopia which had Aksum for its religious capital. In that city is to be found the "*garden of stelae*" among which, lying on the ground in pieces, is a huge obelisk whose length of 35 m exceeded that of the one flaunting its health in Piazza San Giovanni Laterano. The Aksum obelisk had been successfully extracted from the bedrock in the quarry and it was during the process of its erection that it had broken. But this does not trouble the youngster sitting on one of the enormous fragments of this giant, which would have been the largest of all finished obelisks (see Figure 31).

Figure 29 ▲
The unfinished obelisk of Aswan in its quarry.

Figure 30 ▲
The lateral trenches hewn out to free the mass of the obelisk.

Figure 31 ▲
The 35 m broken obelisk at Aksum.

Although the cutting of the stone in the quarry and then its towing along the Nile were extremely difficult operations, they were not the most dangerous: it was during their erection that many obelisks broke because they were too fragile to withstand the flexion they were subjected to during the operation. As in the case of the raised stones of the Celtic period found along the Atlantic seaboard, the obelisk, its base leading the way, was dragged to the top of a mound of the same height as the future obelisk in position (Figure 32). After being hauled across the apex of the mound the obelisk was let down with extreme care until it came to rest on its future pedestal. The operation had to be carried out gently so restraining ropes were attached behind it to prevent it accelerating down the slope. The longer the obelisk the trickier the operation. To avoid excessive overhang towards the rear, ropes were carefully positioned in front to help it pivot more easily. All this merely attenuated the enormous bending stress, but unfortunately the Aswan granite was quite the opposite of Montaigne's view of human reason, which he regarded as "*an instrument of lead and wax, stretching, pliable, and that may be fitted to all byases and squared to all measures*"[26].

Pliny the Elder gives us a glimpse of this delicate operation. He tells us about a king of the eighteenth dynasty who wanted to raise a large obelisk: "*When he was on the point of raising it to the vertical, the King himself became fearful that the devices being used might not be powerful enough to cope with the weight. To better impress on his technicians the peril of the situation, he ordered his own son to be bound to the tip of the obelisk so that their efforts to save him would benefit the stone*".

One can imagine the terror of that child on hearing the first murmurs of strain within the needle to which he was bound. He knew full well what would happen to him if those murmurs grew louder. His only consolation was to read the hieroglyphics that sang the praises of his father the king.

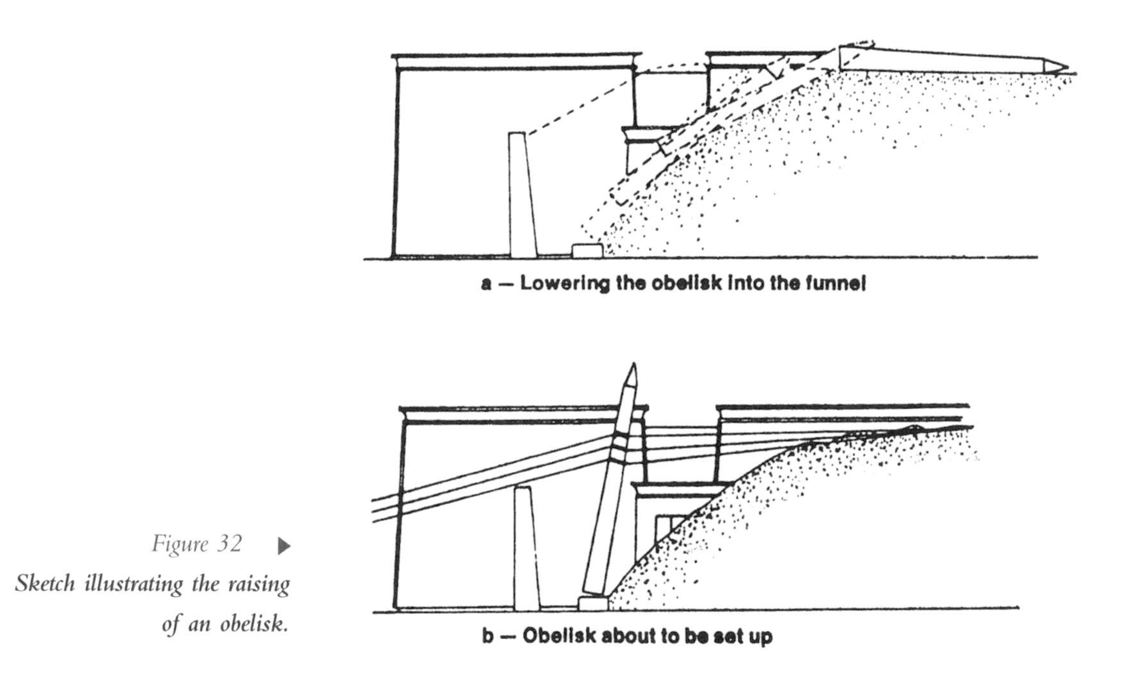

Figure 32 ▶
Sketch illustrating the raising of an obelisk.

[26] Essays of Montaigne, Book II, Chapter 12. Translated by John Florio, 1603.

This curious use of a son calls to mind an example of the strictness of Hammurabi's Code of Laws (1793-1750 BC) in Babylon. This Code drew a distinction between notables, ordinary people and slaves, with each category having distinct rights: if it happened, for instance, that a house built for a notable collapsed and crushed the notable's own son, the notable would then put to death the son of the builder. One version of the principle of an eye for an eye with the potential to decimate families!

By the nineteenth century, on the other hand, the many advances in the science of materials made it possible to calculate the bending capacity of stones and identify the points of strain. It was then a simple matter to erect the obelisks brought from Egypt.

◄ *Figure 33*
A nineteenth century photograph from the Griffith Institute in Oxford showing 'Cleopatra's Needle' at Alexandria about to be shipped to London (photographed by Borgiotti in 1877).

◄ *Figure 34*
Erection of the obelisk in New York's Central Park in October 1880: no sign of the President's son tied to the point!

But that is not the end of the history of obelisks. In 1937 Mussolini, having conquered the ancient realm of the Queen of Sheba and flushed with his victory over Haile Selassie, the "*king of Ethiopian kings*", confiscated the highest of the remaining obelisks in the garden of stelae at Aksum. At 23 m it was a relatively small specimen, a leftover, so to speak. Even so, jealous as he was of the glory of the pharaohs, he had it erected in Rome in front of his Ministry of Africa – now the headquarters of the United Nations Food and Agriculture Organization (FAO). Ethiopia now wants Italy to return the stela: four times, in 1948 in the Treaty of Addis Ababa, in 1956, in 1997 and in 2002, Italy has promised to return it but so far no action has been taken. Tempers are fraying: "*It is not just an Ethiopian problem, it is an African one*", insists Seyoum Mesfin, Ethiopia's minister of foreign affairs, "*this obelisk is the heritage of the black people*".

This modest obelisk, which in fact is more like a stela, has thus become an emblem. But for technical reasons its return is a highly risky business. Broken into four fragments by an earthquake, the Aksum obelisk was repaired in Italy. In addition the journey from the port of Massawa to the highlands of Ethiopia along a road with numerous hairpin bends will present major difficulties.

Figure 35 ▶

Struck by lightning in May 2002, the Aksum obelisk or stela will need to be thoroughly restored before its return. (Photograph Gregorio Borgia, Sipa.)

5

The art of covering space: stone capitals, architraves and vaults

Those who erected obelisks had learned to their cost how difficult it was to tame the laws of equilibrium and avoid subjecting the long stone to bending stress. These problems led to the conception of Egyptian temples, where the columns are placed close together and topped with very wide capitals on the edge of which fairly short lintels are positioned to bridge the gap. In Egypt, and especially at Karnak, there still exists the "forest" of 134 columns supporting capitals in the form of a papyrus that has just blossomed in the sun. The resulting twilight effect was perfect for the intimate atmosphere of the temples and the secrecy of oracles.

◀ *Figure 36*
The capitals of ancient Egyptian columns imitated the form of the papyrus or palm-tree and suggested the vegetation of a marsh.

▲ Figure 37

Note how close together the columns and capitals in the hypostyle hall of the Temple of Horus at Edfu are. (Photograph Patrick de Wilde, Hoa-Qui.)

The Greeks, on the other hand, were lovers of light and the contrasts it produced. In their search for shade and coolness they constructed roofed colonnades. Thin widely spaced columns with smallish capitals supported the lintels, friezes and cornices above. The Greeks went boldly ahead even in areas where earthquakes were common. To make the building more solid, they placed the columns on orthostats positioned symmetrically opposite the lintels above in order to spread the weight. In addition, to improve the monolithic quality of the ensemble, the joints between orthostats and between architraves were reinforced with iron cramps let into the stone and sealed with molten lead.

Figure 38 ▶

Reconstitution of the Attilus stoa in Athens. (Photograph by permission of the American School of Athens.)

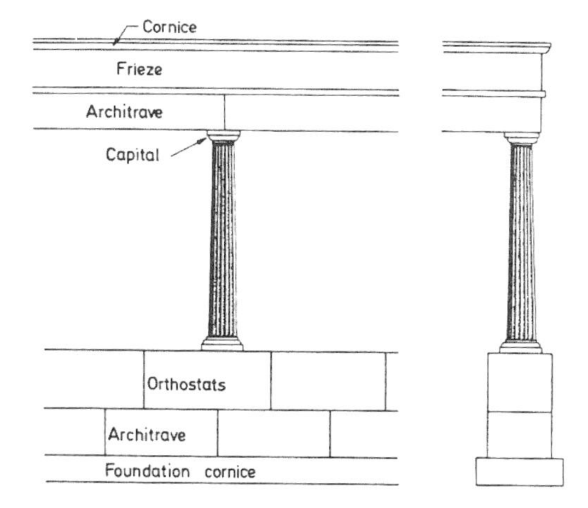

◀ *Figure 39*

A symmetrical structure above and below the columns.

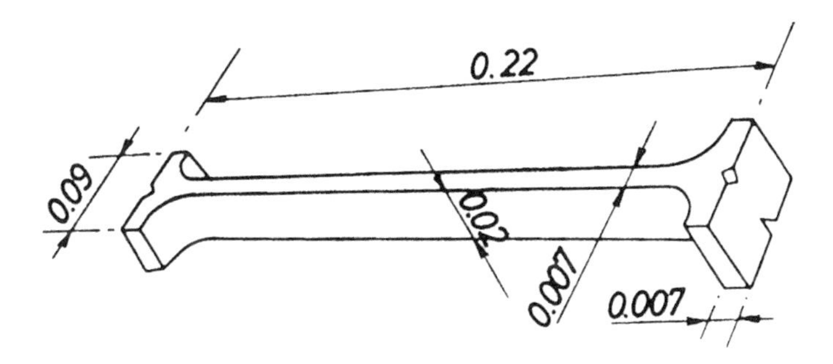

◀ *Figure 40*

Iron cramps were used in both the cornice and the foundation.

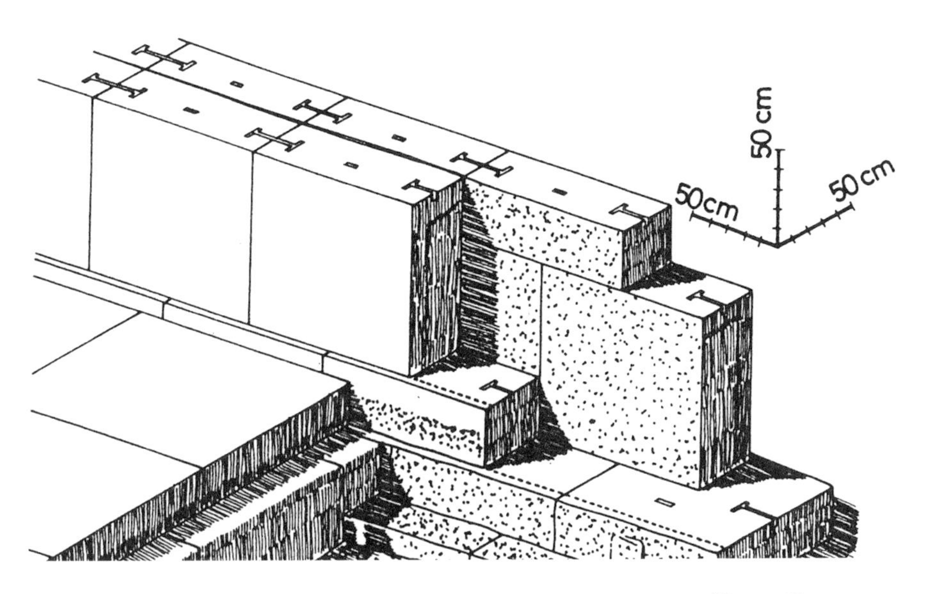

Figure 41 ▲

Detail showing the placing of the orthostats and cramps.

Instead of wide capitals, caryatids were sometimes used with a very small supporting surface, and the distance between them was steadily increased. But experience showed that the boldness in comparison to Egypt had been taken too far: the cramps gave way or rusted and the lintels broke.

Eventually, something will have to be done about the rusting of the cramps: nowadays, for the Parthenon, cramps made of titanium, which does not rust even in marine conditions and has a coefficient of expansion close to that of stone, are being used. But the operation is very expensive.

At the time of the brilliant Republic of Athens, the Greeks had already forgotten the lessons of their forerunners in Mycenae[27] in the Peloponnese a thousand years before. In that city, efforts had already been made to understand the reasons for the rebellion of the mineral kingdom, which seriously threatened lintels and architraves.

The Mycenaeans rightly considered that a lintel suffers because its underside is stressed: it

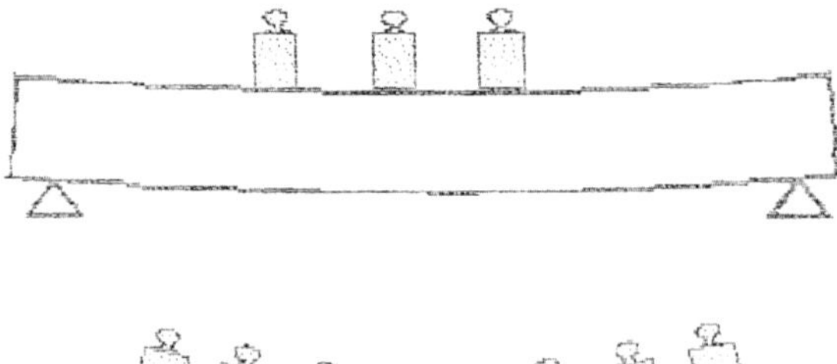

Figure 42 ▶

Deformation of a lintel (top) before its rupture (bottom).

Figure 43 ▶

The Lion Gate at Mycenae.

[27] Mycenae was a flourishing city in around 1600 BC and was the principal city of Greece during the Mycenaean civilization.

bends and, if the burden is too heavy, breaks. It is essential to relieve it in one way or another.

The city was surrounded by an enormous cyclopean wall 17 m high and with an average thickness of 6 m. One entered through the Lion Gate[28], a masterpiece of prodigious lucidity. Above the lintel is a triangular stone of relatively light limestone. In other words, the Mycenaeans had already identified the zone of masonry that would bear down upon the lintel and had reduced its weight. Nowadays, we can make use of scientific studies to show that the shape of that triangle is quite close to the real pattern of stresses. The triangular stone also paid tribute to the art of sculpture for it was decorated with two wild beasts face to face. This motif suggests that the Mycenaeans had grasped the fact that it would be possible to take matters further by eliminating both the triangle and the lintel and leaving the stones on each side to confront each other. That is how they conceived and built the covered stairway that descends to the cistern of Perseus.

◀ *Figure 44*
Mycenae: the vaulted tunnel that descends to the cistern of Perseus.

[28] Lions had long existed in the region.

But that is not all. Outside the citadel lies the Treasury of Atreus, which has one of the most beautiful vaults of the ancient world.

Here the mineral kingdom is treated with great tact: the stones support each other both horizontally and vertically. Roger Caillois would have seen it as a kind of mineral phalanstery[29].

How astonishing it is that such an ancient culture has been able to solve, within such art, three fundamental problems in architecture. The Republic of Athens, with its simplifying classicism enamoured of geometry, would forget Mycenae and its works so redolent of logic and rationality. The Greeks were so sure of themselves that they would persist in their approach to architecture, which was undoubtedly beautiful but dangerous in the long term.

The Greeks had a sense of grandeur and were convinced of their own superiority; they had the impression of dominating the world of "barbarians" that surrounded them. When, towards 450 BC, Herodotus, travelling from Helicarnassus, arrived in Athens, it was proposed that he should visit the Egyptians, the "barbarians of the south". He went to Egypt and spoke with the priests of the "Houses of Life" that juxtaposed the great temples. He was dumbfounded by their culture and the richness of their archives. After his return, Plato, in one of his books, put these words into the mouth of one of those Egyptian priests: "*You others, you Greeks, are children for you have no ancient history whilst here nothing of greatness or of beauty is done without the memory of it being preserved in our temples*". For the Egyptians, the barbarians were the Greeks: and perhaps they were right, for had not the Greeks totally lost all recollection of Mycenae, the jewel of their architecture, now almost buried under a shroud of sand?

The early Romans, with their horror of sophistication, remembered only the vault. The lofty rectangles of the Greek porticos gave way to powerful semi-circular vaults in which the lines of force are absorbed by a superabundant stonework. They were as solid as the robust Roman legions of that time.

The celebrated treatise of Vitruvius, *De Re Architectura*, which was published in the first century BC, shows that there existed other currents of thought on the subject of architecture in Rome. Material proof is provided by the Roman aqueduct that crosses the river Gard in the south of France. Its construction was ordered by Agrippa, a general with the full confidence of Augustus. It was built in stone during the first century AD. Its three levels give it a very elegant appearance. The most important level is the lowest one with its six large arches that support a second level of eleven large arches and a third level of thirty-five small ones. The whole structure exemplifies Vitruvius' recommendations in Book I, Chapter 2, of his treatise: "*Architecture depends on Order, Eurythmy, Symmetry, Propriety and Economy.... Order gives due measure to the members of a work considered separately. Eurythmy is beauty and fitness in the adjustments of the members.*"

It is Marcel Pagnol, speaking of his grandfather, that we should call to mind to evoke this great

◄ *Figure 45*

Mycenae: the vaulted ceiling of the Treasury of Atreus. (Photograph Dagli-Horti.)

[29] A new type of community imagined by the French sociologist Charles Fournier (1772-1837) in the 1830s. A large group of persons would live in a community with common activities, interests and aims.

▲ *Figure 46*

The Roman aqueduct at Caesarea, capital of the imperial province of Judea (repaired in the second century AD). (Photograph Magnum.)

Figure 47 ▶

The Roman aqueduct crossing the river Gard with its close-fitting voussoirs. Painting by Hubert Robert, 1787 (Louvre Museum). (Photograph Gérard Blot, RMN.)

alignment of arches. "*Whenever he had a day free, that is five or six times a year, he would take the whole family for a picnic lunch on the grass about fifty metres from the Pont du Gard. While my grandmother got the meal ready and the children dived into the river, he would climb up on to the floors of the bridge, take measurements, examine joints, draw detailed cross-sections and caress the stones*". At midday, under the gaze of this old man, I imagine the family group spread out on the grass for their meal in a graceful curve like one of the lower arches. "*After lunch he would sit on the grass with the family and contemplate until evening the two-thousand-year-old masterpiece*"[30].

The stone arch, logo of a Europe in the making

The illustrations on Euro banknotes give pride of place to stone arches and bridges. The intention was to symbolize communication between peoples. The designer was perhaps thinking of the painting by Joseph Vernet entitled "Construction of a highway", with the arches of a bridge under construction in the background (see Figure 48). The banknotes remind us of the long road ahead to build a united Europe. They even offer, as their face value increases, a history of the design and structure of stone bridges. It was a bright idea but not flawless in its execution.

Figure 48 ▲

Construction of a highway. Painting by Joseph Vernet, 1774 (Louvre Museum). (Photograph Franck Raux, RMN.)

[30] Marcel Pagnol, *La gloire de mon père*.

On the modest five-euro note, for example, we see an aqueduct (see Figure 49). What is shocking is the designer's total ignorance of engineering: his representation would have greatly irritated the builder of the Lion Gate, the erudite Leonardo da Vinci or the self-educated grandfather of Marcel Pagnol, all of whom were able to sense the lines of force. On the right hand side of this poor man's banknote we see a heavy pillar of the second level pressing down on the keystone of a large arch below. Leonardo da Vinci, in his poetic but frequently obscure language, said that a vault had "*due debolleze*", i.e. two weak points: its crown and its abutments. The latter were able to strengthen the arch *only if the forces pressing on the crown were weak*, which is certainly not the case in this sorry drawing; if, on the other hand, the abutments are strong enough to resist overturning, as in the case of the construction of the aqueduct since they are supported by the neighbouring arches, "*si convertano en una fortellezza*", they become a source of strength. Exactly the same applies to the haunches of the vault, marked '2' in Figure 50, which are highly vulnerable if not adequately reinforced.

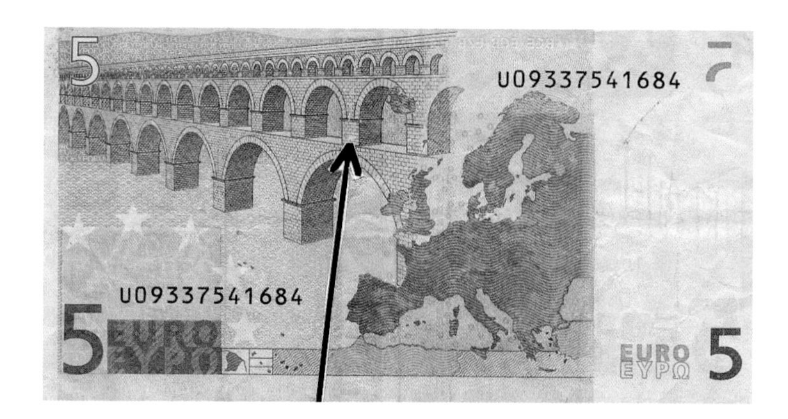

Figure 49 ▶
The five-euro banknote and its major error in design.

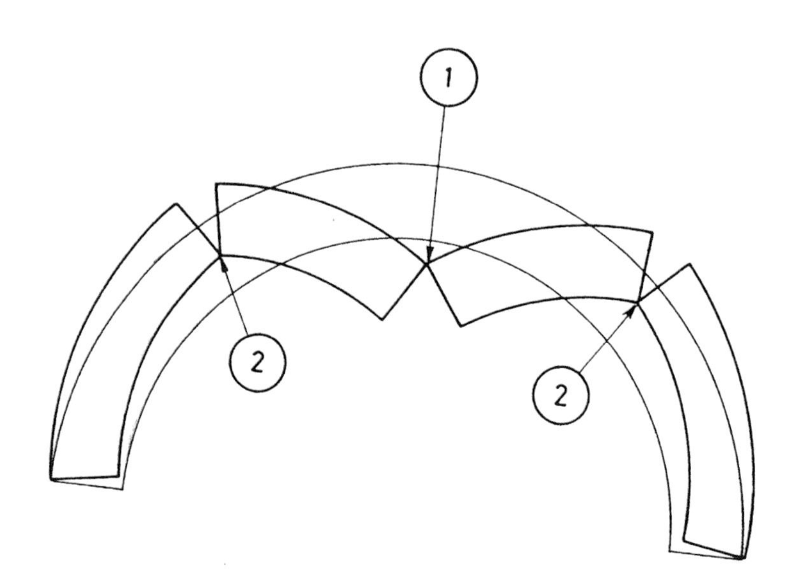

Figure 50 ▶
Sketch illustrating the views on arches of Leonardo da Vinci.

Death of the stone vault because of excessive demands; the last great masonry arches

The green 100-euro banknote is the last to pay tribute to the stone arch. Two arches are reflected in the water. On the larger denominations of 200 and 500 euros, the stone arches and all they have contributed to human life over the centuries in the bridging of powerful rivers are forgotten. And yet bridges have often been the heroes of local history. In Orleans, people have always admired the Pont Royal, which had withstood, in the person of Madame de Pompadour, mistress of Louis XV, "*the heaviest burden in the kingdom*" as well as the three enormous floods of the Loire river in 1846, 1856 and 1866.

The building of bridges was often a risky business. In another case concerning the Loire, the King had ordered the construction of a bridge across the river at Moulins. But this bridge was swept away by the river in flood and the unfortunate gentleman engineer in charge of its construction had to reply to an enquiry from the King: "*Sire, my bridge is no longer at Moulins but is already at Sully and will be tomorrow at Orleans...*".

Figure 51.a
The last large bridge in masonry: the Pont Adolphe in Luxembourg, built in 1903 with a span of 84 metres.

Figure 51.b
A recent postage stamp showing the bridge.

All such works had their moments of glory or despair. Record after record was broken in length of span, the structure was lightened and embellished, and little by little the efforts demanded of the poor stones became impossible to sustain. As soon as the formwork that supported the arch during its construction was removed the half-arch on one side confronted its opposite number like two rams fighting over territory or over a female. All the stones were wedged tightly together; if the area of the keystone dipped slightly it would cause a wide crack to appear on its underside – the arch lost its gracefulness and became fragile.

The stone arch died in 1903 in the Grand Duchy of Luxembourg. The task there was to bridge a deep ravine by means of a free-standing arch with a span of 84 metres. The Grand Duke preferred the warm colours of stone to the coldness of concrete, which was being used almost everywhere, reinforced with steel rods to offset the weaknesses of stone in bending and in traction. A page was being turned: Paul Séjourné was, along with Gustave Eiffel, the most highly reputed engineer of his time. He accepted the offer of the Grand Duke and succeeded completely in his wager: his bridge, the Pont Adolphe, still holds the record for the span of a stone bridge. The stones work together perfectly. It is the largest stone arch in the world and Paul Séjourné was the last great artist of masonry bridges.

6

Inside the Great Pyramid

While recognizing the immensity of Cheops' genius[31] I hold that his authoritarian attitudes and his hubris led him to commit a serious mistake in the construction of his exceptional pyramid.

Myths refuse to die; there will always be people with a boundless admiration for the Egyptians who built the Great Pyramid. They see the monument as the superposition of magnificent stones cut to the tenth of a millimetre on each of their six faces. Others, without having ever been inside the pyramid, go further and claim that these stone parallelepipeds, like giant sugar lumps, were made of concrete poured into moulds[32]. What a mistake! The Egyptians of the time were far from knowing everything and had no idea about recent techniques of cast concrete. Slender in build and with a life expectancy of only about twenty years, the Egyptians could not possibly cast and then drag some five million tonnes of stones. One has only to see a few of the corridors inside the pyramid to find evidence of the inaccuracy of these assertions. On the contrary, we find two quite different kinds of masonry side by side, a masonry of the rich and a masonry of the poor. The latter unexpectedly ends up by causing damage to the former.

Each major stone monument embodies a kind of wager on the part of its builder: Cheops wanted the architecture of the Great Pyramid to represent Egyptian society and its profound contrasts. To this end the pharaoh reserved, on the northern side of the median half-plane, a narrow zone for access to his chamber in the centre of the pyramid and decided that this small area would symbolize himself. So he built it using the noblest materials in his kingdom – the granite of Aswan, which had to be transported over a thousand kilometres and, under it, limestone of excellent quality quarried on the opposite bank of the Nile.

These "*nobles*" are surrounded by the people, "*the human cattle*", as the subjects of the pharaoh are called in the texts, who are represented by more modest materials – limestone stuffed with nummulites, tiny organisms that lived some fifty million years ago and were sealed in their tombs after the disappearance of the tertiary seas. They are like those stones of which Roger Caillois speaks so well at the beginning of his essay on stones: "*I speak of stones that have always lain outside or which sleep*

---

[31] See Kerisel, J., *Génie et démesure d'un pharaon: Khéops*, Paris, Stock, 1996 and 2001.
[32] Odd idea for a country without lime: the binder used for the pyramid was plaster.

in their shelter or in the dark of their lode. They are of no interest to the artist, the archaeologist or the diamond-cutter. Nobody uses them to build palaces, sculpt statues or shape gems; but dykes and ramparts ... constructions without honour or reverence, testify to their use".

Cheops founds such stones only a few hundred metres from his pyramid and, probably without seeking "*the pardon of the lord of the hills and valleys*", created a huge scar in the hillside that is still visible today.

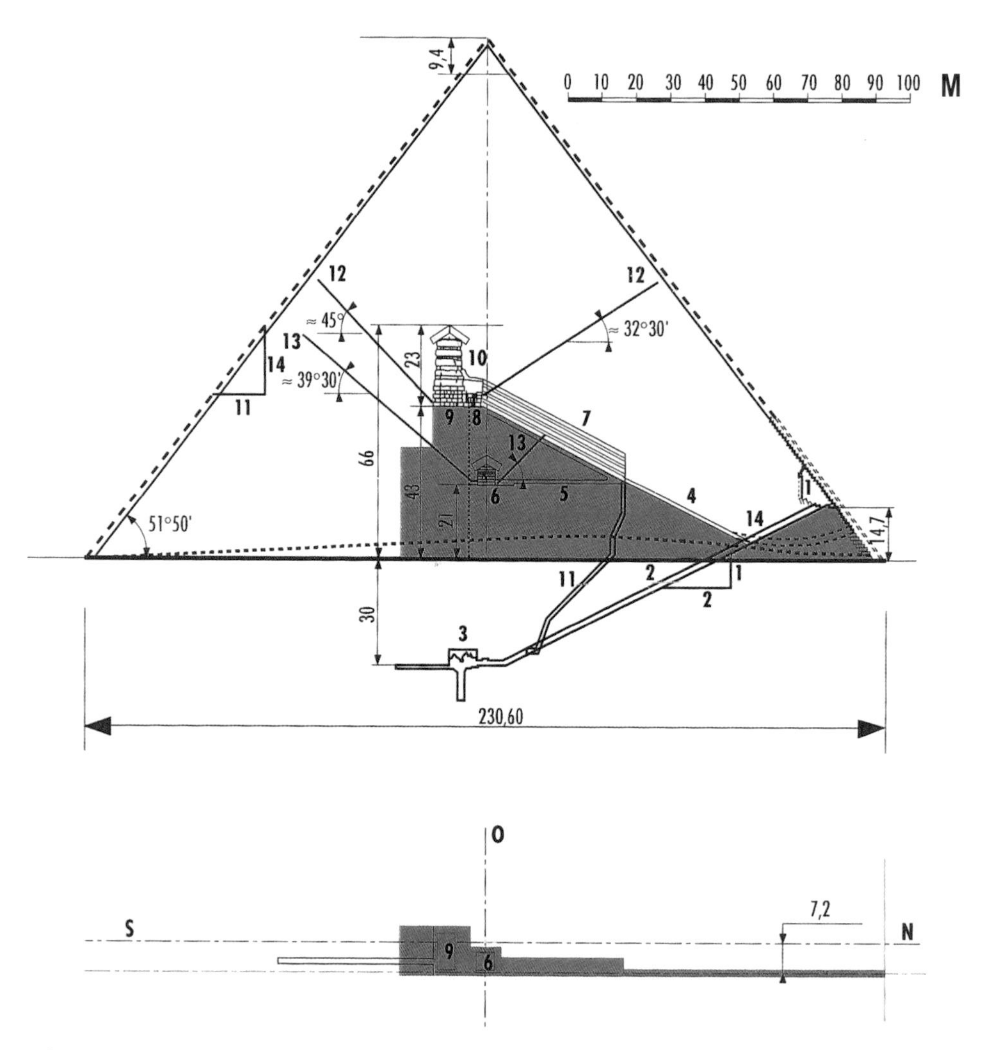

▲ *Figure 52*

The Great Pyramid. Vertical section seen from the east and axis of the corridors seen from above. A tiny space was reserved for the royal architecture in the northern half-plane of the North-South cross-section (red zone in the vertical and horizontal sections) 1. Entrance 2. Descending corridor 3. Underground chamber 4. Ascending corridor 5. Horizontal corridor 6. Queen's Chamber 7. Grand Gallery 8. Antechamber 9. King's Chamber 10. Upper compartments 11. Shafts and service corridors 12. Vents for the King's Chamber 13. Vents for the Queen's chamber 14. System to block the ascending corridor.

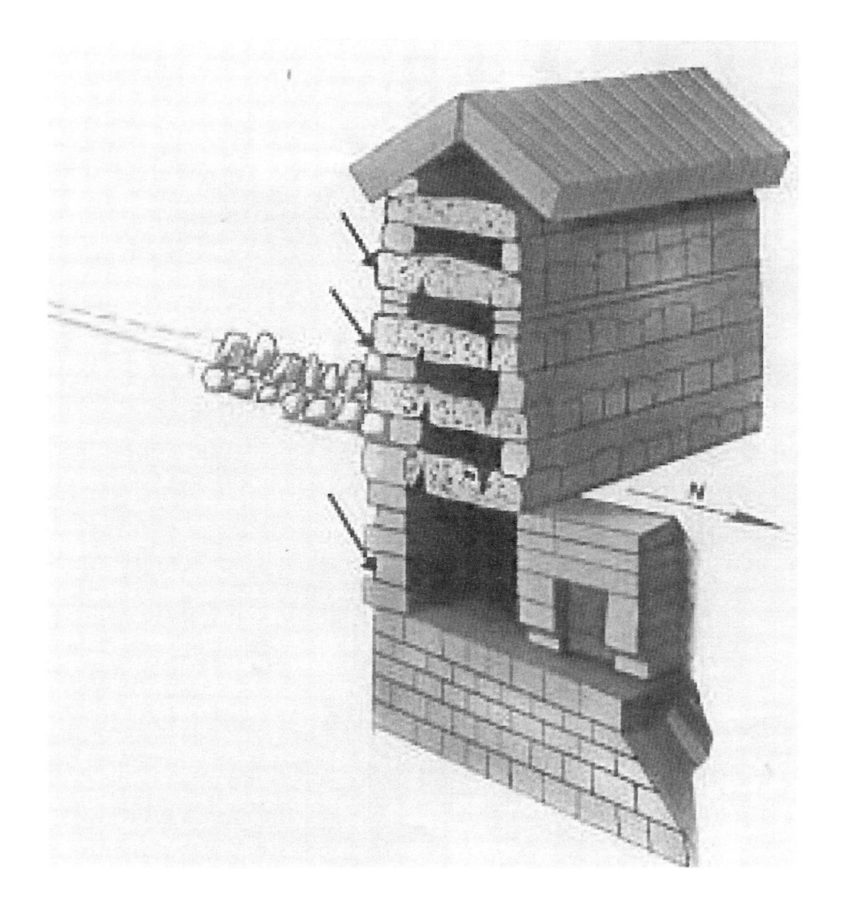

◀ *Figure 53*
Interaction of two types of masonry with differing compressibility along the south wall of the King's Chamber, causing distortion and rupture of the granite beams (underside at their southern extremity and upper side at the northern one).

This arrogant composition sought through its symbolism to emphasize the pre-eminence of the pharaoh. The chamber was positioned on a powerful base, like a sheet of steel, with an extension northwards for access and in the midst of but quite distinct from a mass of more compressible materials, a community of humble stones roughly stacked together. It was like an oak tree surrounded by reeds.

Gradually, as the pyramid increased in height, the 'humble' masonry, comprised of roughly shaped stone blocks clumsily embedded in a gypsum mortar, was necessarily vulnerable to compression. As the pyramid rose and the weight of the structure became enormous, this masonry was squeezed and pressed down on the pointed roof, which gave way; the upper compartments of the King's Chamber were then overburdened with the result that the south wall was subjected to a massive downward friction that tipped the chamber itself in that direction. The distortion was verified by Sir William Flinders Petrie and rechecked several times by the author[33].

In their daily lives the human cattle had no choice but to obey the commands of the tyrant pharaoh; in this case, however, the laws of mechanics ensured that their masonry counterparts had the consolation of revenge: subjected to flexion, the luxurious granite beams in the ceilings and floors, perfectly adjusted to fit the tops of the walls, ruptured with an enormous cracking sound (Figure 53).

[33] At certain points near its centre, the south wall has dropped 5.5 cm in relation to the north wall opposite.

The beams forming the roof and floor were nearly two metres thick and had a cross-section of almost two square metres: their rupture released an enormous energy and it is easy to imagine the noise made by such a hard and rigid granite. They exploded like a gunshot, but it was also a political explosion. The news of the disaster was amplified by those working on the pyramid, who fled in terror, and it spread throughout the country among a harassed people. The open splits in the handsome pink granite of the King's Chamber allow us to imagine the panic and the anguish of the pharaoh himself as, cursed by the priesthood for his arrogance, he hastily finished his enormous pyramid. The pharaoh was cursed and betrayed by his arrant pride and his stubbornness into making a serious mistake.

For all thinking people of the time, a burial chamber evoked the quest for perfection: believing that the pharaoh was inspired by the gods, who would receive his mortal remains, the people had willingly accepted an exhausting challenge, which consisted in hauling up to the level of the compartments above the King's Chamber, which were roughly a hundred metres above the level of the Nile, about sixty granite beams each weighing some sixty tonnes. It was a superhuman effort that ended in a painful and humiliating fiasco: the pharaoh was meant to be infallible and his pyramid under the protection of the gods. The people discovered that the pharaoh had made a mistake, and it was a mistake that came after a long period of suffering.

Those great stones have a memory: dragged up endless ramps, they heard the groans of the men hauling on the ropes, their heavy breathing and the cracking of the whips of the overseers, before being wounded in their own flesh. They are eternal witnesses to the arrogance and hubris of the pharaoh and the malediction of the gods. After the death of the pharaoh, the people took revenge by smashing all the statues of him.

A magnificent example of the relation between a social fact and the mineral kingdom which would have delighted Roger Caillois, who would have observed that, in this combat between David and Goliath, it was David who won. It also shows that ranks do not exist in the mineral kingdom and that the stones never used to build palaces are just as precious as the others.

Herodotus

We must not leave the Great Pyramid without mentioning the testimony of Herodotus regarding the true burial chamber of the pharaoh. The pharaoh had undoubtedly made the error of judgement we have just described, but his genius shines in all other aspects of the project and particularly in the arrangements made for his tomb. Herodotus was absolutely right in affirming that Cheops lay in an underground chamber beneath his pyramid, with access by an underground canal brought into service on the death of the pharaoh. It would appear that Herodotus was more knowledgeable than present-day Egyptologists on matters concerning the water table, diversion canals and load-bearing canals. Few Egyptologists believe in this interpretation. We refer the reader to two studies written on the subject[34].

[34] Kerisel, J., 2002: 'The tomb of Cheops and the testimony of Herodotus', in *Discussions in Egyptology*, 53, pp. 47-55. See also Kerisel, J., 2003; "Le tombeau de Khéops et une vérification peu coûteuse", revue *Travaux*, No. 796, April 2003, Paris, pp. 68-76.

After the accident with the King's Chamber, the pharaoh hastened to finish his pyramid. After his death, the people would smash all the statues of their king[35]. The pharaoh, fearing this and anxious that it would affect his second life, frequently consulted the stonemasons of death, who knew all the secrets of the aqueduct, and the priests who would pilot the funeral bark.

[35] Cheops is one of the few pharaohs of whom no statues have been found.

7

"*Tutti fuori, niente dietro*" all for show, nothing behind

On a fine day in 1902, the magnificent bell-tower built in 1173 collapsed on to the Piazza San Marco in Venice. It had acted as a lighthouse to guide sailors in the lagoon. In medieval times, however, it had served a less charitable purpose: condemned persons were shut into cages that were then hauled halfway up the inside of the tower. The tower had survived a number of vicissitudes, including earthquakes in the sixteenth century, up until 14 July 1902, when it finally gave up the ghost after two days of agony.

The accident made a big impact throughout the world, especially among the numerous admirers of the city of the Doges. Leading scientists were asked for their views. The campanile had not been shaken by a recent earthquake and was perfectly vertical. The enquiry, as is often the case in such matters, simply pointed to the great age of the tower and suggested that this was frequently the fate of old buildings. As the famous surgeon Paul Broca once remarked, "*Life is the conjunction of all the forces that fight against death*". It was as straightforward as that: the bell-tower had simply died of old age!

In 1989, unfortunately, the civic tower of Pavia collapsed in a few seconds, killing four persons. It had been, like the Venice campanile, perfectly vertical. Not the slightest sign of an earthquake had been detected. The incident made much less noise but was nevertheless astonishing. Was there such a thing as an age-limit for bell-towers?

At this time, the Tower of Pisa was beginning to lean more and more and was causing grave concern. An international competition had been launched after an abundant description of the venerable monument in three large-format luxurious volumes published in 1971. Numerous erudite articles left the impression that the superstructure was composed of rectangular blocks of marble perfectly adjusted one on top of the other, the marble coming from San Giuliano or the Apuan Alps, where magnificent quarries are to be found not far from the tower in a region with the most extensive seams of marble in Europe.

The splendour and the abundance of this marble seemed to rule out the possibility of cheating by the builders of a tower raised to the glory of the world's leading maritime power at the time.

▲ *Figure 54*
The marble quarries of Carrara in the Apuan Alps. (Photograph by the author.)

Many nations responded to the appeal for contenders and about forty projects were entered for the competition, whose purpose was to cure the tower of its sick foundations[36]. True to the nature of bureaucracy, the competition failed to evolve into effective action, and no explanation has ever been given for this.

Someone had the idea of conducting an autopsy on the ruins of the tower of Pavia. In the enormous mass of rubble, only a small quantity of marble fragments of the kind that had decorated the facade was found; the ruins were almost entirely composed of mortar and small stones and even "*arena sin calce*" (sand without lime). In other words, the facade had consisted of simply a thin casing of marble; it was an outstanding example of the old Italian saying "*tutto fuori, niente dietro*" (all for show, nothing behind). Historians of art then confessed that that was a common practice for buildings in the middle ages and that it was quite possible that all the towers that had embodied the power of the young Italian republics in the twelfth and thirteenth centuries, including the Tower of Pisa, had been built in that way.

The *tutto fuori niente dietro* approach was clearly a way of saving money, but it was a marriage

[36] The angle at which it leans, already close to 5°30', was increasing alarmingly.

against nature: you can't hitch up together a carthorse and a racehorse. The outer casing, rigid as the glass of a lamp, supported all the weight and cracked or shattered. This desire to show off is at the root of many disasters in the history of architecture and reminded me of some thoughts of the grandfather of Marcel Pagnol[37], who was a mason and expert dresser of stones. Here is what he said of ordinary masons, of whom he did not think highly: "*As for us, we used to build the walls with dressed stone, that is to say stones that fitted exactly into each other with tenon and mortise, with joggle joints, with dovetails and with joggled and wedged scarfs ... Naturally we also poured molten led into the slots to prevent slippage... Whilst the masons took stones as they came and plugged the gaps with wadges of mortar*[38]...". And he finished this severe and intelligent judgement by adding: "*A mason drowns stones in mortar to hide them because he does not know how to dress them*".

In the case of Pisa, such criticism might appear exaggerated, for the ensemble composed of the Cathedral, the Tower and the Baptistery give such a strong impression of being a miracle in marble (Figure 55). These buildings were the jewels of the Piazza dei Miracoli; the immaculate whiteness of the marble tower had even inspired the very sober art critic, Hippolyte Taine, after his trip to Italy in 1864, to call it: "*This beautiful cadaver of glistening marble*". It was true: the tower has always been an illusion. But no cadaver has ever caused so many problems!

The construction of the tower began towards the end of the twelfth century, in 1174, long after the cathedral. It was in fact founded on ancient alluviums of the Arno in which patches of clay alternated irregularly with sand. There was much more sand in the northern part of the site than in the southern one. Contrary to what the Old Testament says, it is wiser to build on sand, as the fifteen thousand tonnes of the tower confirm: when the construction had arrived at a third of the planned height, the tower was found to be leaning so far towards the south, which was poor in sand and rich in clay, that work had to be suspended. Ominous omen for the Republic of the Sea, as Pisa was called! Work recommenced a century later, in 1275, and the tower reached two thirds of the planned height. The angle at which it leaned increased but, even more serious, on 6 August 1284, the Genoese navy, in a sea-battle at Meloria, totally defeated the Republic of the Sea and annihilated the Pisan fleet.

This disaster marked the end of the splendour of Pisa and its profitable trade, but the Pisans nevertheless persevered in their efforts to complete the Piazza dei Miracoli, conceived at the time when it was all-powerful. In 1360, the last floors and the belfry were constructed.

Galileo, born in Pisa in 1564, used the tower for his first experiments with falling objects.

But the beautiful cadaver of which Hippolyte Taine speaks continued to misbehave. Furthermore, judging from what had happened at Pavia, the superstructure had worried the last team of builders as much as the foundations: it was noted that the thin plates of marble cladding were not properly positioned one above the other and created an overhang in certain places. Behind them the filling was curiously poor in quality, with many cavities surrounded by masonry containing a high proportion of mortar – two problems instead of one.

---

[37] Op. cit., p.13.

[38] In the case of the Tower of Pisa, the voids left between the large stones made up about 30% of the total volume.

Figure 55 ▶
The Piazza dei Miracoli in Pisa.

V ≅ 142 MN, M ≅ 327 MNm, e ≅ 2.3 m
Situation in year 1990

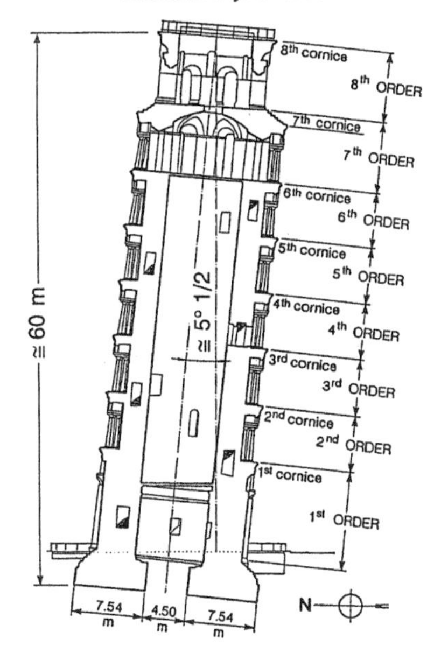

Figure 56 ▶
Vertical section of the Tower of Pisa in 1990, showing the eight orders and the cornices (V = weight, M = moment of overturning, e = M/V). The Tower leans towards the south at an angle of nearly 20 000 seconds of arc, or about 5°30'.

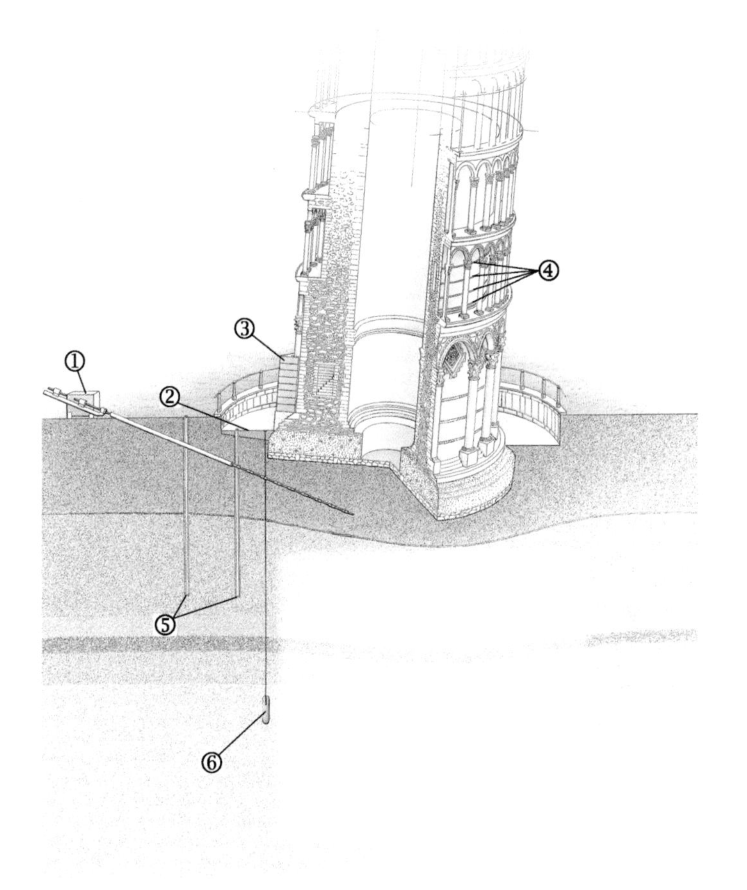

Figure 57

Tower of Pisa : cross-section seen from the west. 1. Drilling to extract earth 2. Concrete ring 3. Lead weights 4. Hooping with steel cables 5. Electrodes for extraction of water by electro-osmosis (project) 6. Cable to anchor the concrete ring to a deep sand layer.
(Photograph by permission of "Pour la Science".)

The angle at which the tower leaned varied so much that it quickly became necessary to abandon the plumb line and metric system in favour of degrees, minutes and seconds of arc, with the latter soon becoming the standard unit of measurement. The angle was nearly 20 000 seconds of arc (roughly 5° 30') at the beginning of the present century. It had been increasing by an average of 4 to 5 seconds per year throughout the twentieth century. During the night of 11 September 1995, however, the angle increased by that amount in a few hours.

Figure 57, seen from the north, shows some of the methods envisaged for righting the Tower: loading with lead ingots (a temporary measure); a ring of concrete anchored deep in the soil (quickly abandoned because of serious incidents); electro-osmosis for the extraction of water (abandoned); drilling at an angle to extract the sandy earth (method used in the end with great success under the general direction of Professor John Burland of Imperial College, London). All honour to the three geotechnical engineers who saved this graceful tower; thanks to them, the angle of inclination

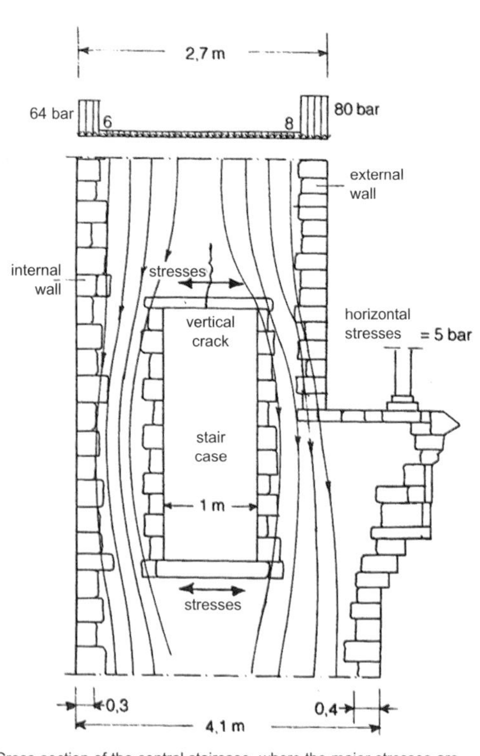

Cross-section of the central staircase, where the major stresses are concentrated [BURLAND et al., 1993].

▲ *Figure 58*

First Cornice of the Tower of Pisa, south side, level with the internal spiral staircase: stresses have considerably increased, with a tendency to extension, leading to rupture of the columns (reinforced with iron binding hoops) and to the need to circle the body of the tower with steel cables at the level of the first order (see Figure 57).

had already been reduced to under 19 000 seconds of arc by February 2001. The method used worked wonderfully well. Indeed, it became necessary to ease up a bit: the local authorities have – understandably – decided that the tower must continue to lean. Otherwise, as Antonio Paolucci, former minister of cultural property, observed, it would kill the myth.

In other words, the problem of the tower's foundations has now been solved and its inclination could in the future be reduced as much as desired; but the second problem remains worrying. In spite of subtle injections and steel hoops to strengthen the cracked columns with a corset of stainless steel, the Tower is more vulnerable to earthquakes than in the past. In the long term, the Tower will always remain the slave of Pisa's dreams of grandeur.

The Pantheon in Paris: Soufllot's stupid wager

Before the French Revolution, Louis XV dreamed of consolidating the monarchy by launching an ambitious programme of building. In 1744, after recovering from a very serious illness thanks, he thought, to the intercession of Saint Genevieve, he vowed to dedicate a prestigious building to her. At the time, the Director of the Royal Buildings was the Marquis de Marigny, brother of the famous Marquise de Pompadour. In 1755, he entrusted his protégé Jacques-Germain Soufflot with the task of building, on the Montagne Sainte-Geneviève that overlooked Paris, the great church that the monarch desired.

This monument, begun in 1757[39], in danger of collapse in 1796 and completed in 1807, provides a good example of many architectural errors. A sense of grandeur is mingled with incompetence, particularly regarding the resistance of the stones, which were tortured by excessive demands until they cracked in many places. It was the same story as the Tower of Pisa but much worse. A subject of bitter disagreement between architects and engineers at the end of the age of enlightenment, the Pantheon engendered passionate debate, insults and even some dirty tricks.

In its architecture every style is represented: it is Greek in the purity of certains forms and Gothic for the lightness of its supports. Its vaguely ambiguous and composite style is explained by the fact that it was sometimes seen as a church and sometimes as a patriotic temple. It is a building that is still dangerous on account of falling stones, costly in upkeep and relatively little visited on the whole because of long periods of closure for repairs.

Jacques Soufflot, who undoubtedly had a sense of composition, wanted to top the church with a large dome, the third largest in the world after those of Saint Peter's in Rome and Saint Paul's in London, so that the patron saint of Paris, Saint Genevieve, could overlook in all her beauty the city she had saved. The cupola was threefold, ie three skins nested one inside the other, and very heavy: each of the four 38-metre pillars bearing the drum of the cupola had to support no less than 3 100 tonnes[40]. He ended up by designing the section of those pillars in the form of isosceles triangles at each corner of a square, with columns inserted into each point of the triangles (Figure 60).

The slenderness of the pillars did not escape the attention of a certain Pierre Patte, who had just visited Rome. In 1770, happy to pick a bone with one of the king's favourites, he published a lampoon in Amsterdam whose title in French translates as follows: "*Memoir on the construction of the cupola intended to crown the new church of St Genevieve in Paris*". Its purpose was to show that the pillars already realized and intended to bear the said cupola were not big enough to allow the raising of such a work with solidity. He added that the memoir was addressed "*to all learned societies, to engineers, to architects and to all with knowledge of matters of construction*". The tone was particularly wounding for Soufflot.

The latter, a mediocre engineer with grandiose ideas, had no wish to admit the blunders he had committed and never perceived the serious consequences they would have. Some illustrious

[39] The first stone of the building was laid by Louis XV at a solemn ceremony on 6 September 1764, but work had been in progress on the foundations since 1757.
[40] According to J. Rondelet, Vol. III, pp. 180-181, the initial project submitted to the king on 2 March 1757 was even more ambitious. The twelve columns were finally grouped by threes because of Patte's criticisms and despite the reassurance of Emiland Gauthey.

scientists, including Charles Augustin Coulomb[41], offered their services at the Académie Royale, but Soufflot paid no attention and did not seek advice. It should be stressed that the Académie Royale, under the leadership of Marcellin Berthelot, Auguste Comte and others, had remained old-fashioned in its views and had a lot of leeway to make up.

The weight-bearing surface of each pillar was only about 15 m^2 for a burden of 3 100 tonnes; assuming that this weight was evenly spread over the entire surface of the pillar, the pressure on the floor would amount to some 200 tonnes per square metre, far from negligible but nevertheless acceptable.

But the filling inside the columns was of very poor quality – a soup of mortar - and from the very beginning could not do its share of the work of bearing the weight of the cupola. The weight-bearing surface of the facing stones, which were 30 cm thick (see hatching in the left-hand diagram of Figure 60), was only a tenth of the surface of the pillar, which greatly increased the pressure. Much more serious was the fact that the stone used for the facing, which came from Bagneux, a suburb to

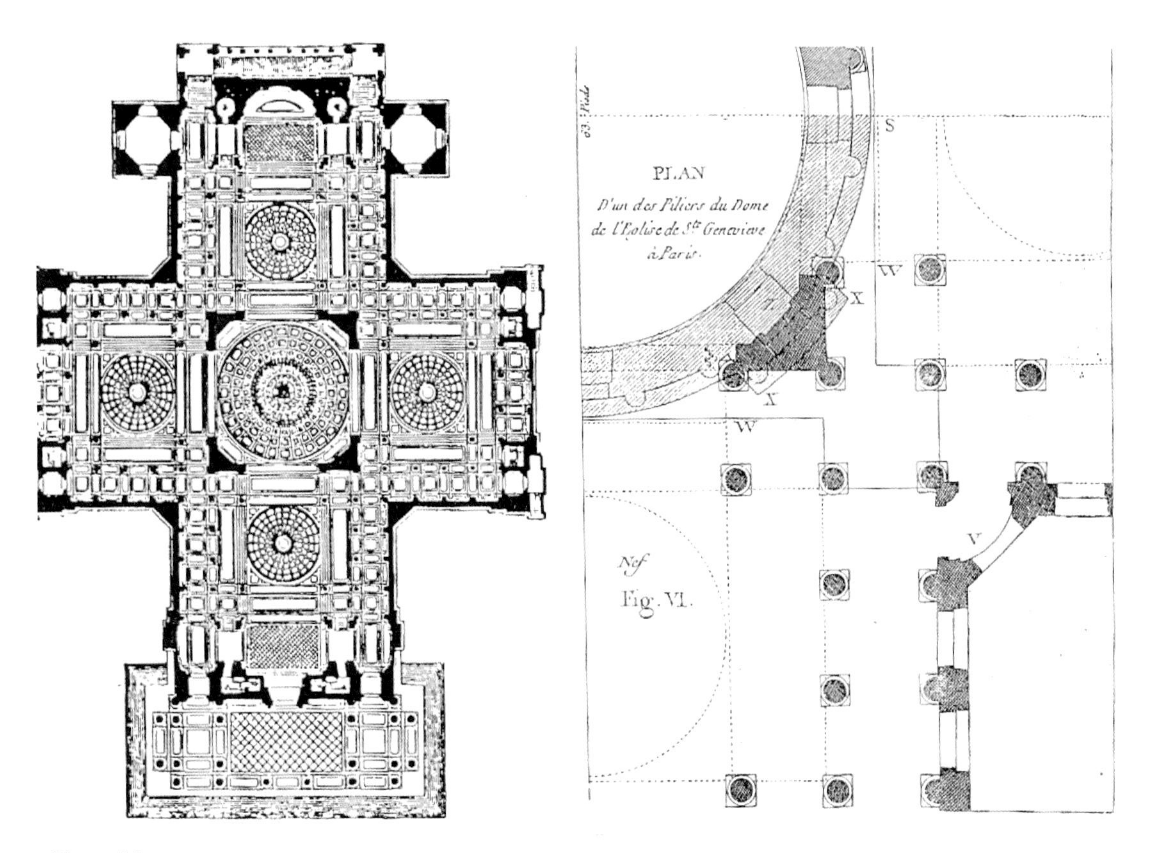

▲ *Figure 59*

Plan of the Pantheon showing the position of the four pillars. One of them is shown in detail in the drawing on the right.

[41] *Charles Augustin Coulomb, Bicentenaire de son essai présenté à l'Académie Royale en 1773*, by Jean Kerisel, Paris, Presses de l'Ecole des Ponts et Chaussées.

the south of Paris, was not of top quality. Here it is worth quoting the shrewd comment of Eugène Viollet-le-Duc, the famous nineteenth century architect: "*What is odd is the use of poor quality materials with the intention of imitating constructions achieved with the aid of very powerful means*".

Worse still, on the eve of the Revolution the royal treasury was empty. Because the builders were not being properly paid, they cheated: they refused to dress the entire width of the facing stones, contenting themselves with less than half (Figure 60, right-hand diagram), thus increasing the theoretical pressure a further twenty times. This was the average pressure, since the dressing of the weight-bearing surfaces of the facing stones had been given out to be done as "piecework", with the result that the stones were poorly trued up, with dull edges, in the opinion of artisans employed by Jean-Baptiste Rondelet, who began working for Soufflot in 1769.

As a result, owing to the dissymmetry between the front and the back, the facing stones did not sit properly: to make sure they were correctly adjusted, the builder inserted oak wedges between which mortar was forced in with a trowel. But the shrinkage of the mortar quickly rendered the attempt to spread the weight ineffectual. In other words, the cupola rested on isolated pieces of oak and the enormous pressure snapped off fragments from the facing stones (Figure 61).

The necessary money should have been spent on the complete dressing of the stones, without skew surfaces or dull edges, by smoothing them with sandstone: this is what the Romans and before them the Egyptians did, showing that they knew how to spread the load the stones had to bear without injuring them. The French Pantheon represents a step backwards in the art of building.

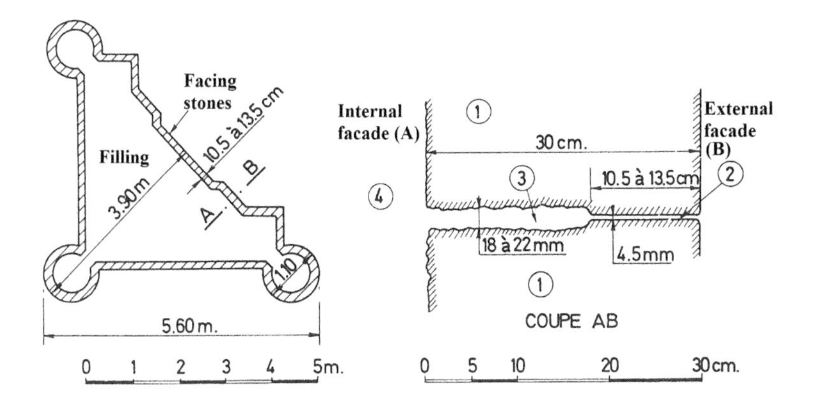

◀ *Figure 60*

Dimensions of a pillar supporting the dome, showing the proportion of mortar filling behind the facing stones, and vertical cross-section of a joint. In the section AB, on the right, 1 = facing stone, 2 = dressed joint, 3 = mortar filling in undressed joint, 4 = core filling of low quality mortar.

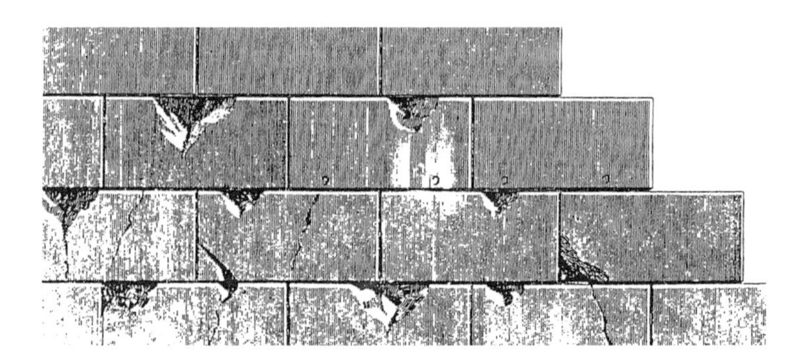

◀ *Figure 61*

A few examples of damaged facing stones.

These stones of the Pantheon, which have suffered so much from the incompetence of human beings, have first-hand knowledge of man's thirst for glory and spinelessness. It is hard to imagine that an architect enjoying the favours of the King should show himself to be so blind and so unaware. To put his mind at rest, Soufflot turned to a few reputed scientists who like himself were in the King's good books.

Emiland-Marie Gauthey (1732-1806), who was both an engineer and an architect, rushed to his rescue in 1771 by demonstrating that the dimensions of the pillars of the Pantheon were in proportion to those of the churches he had himself constructed in the Saône-et-Loire département. It was a worthless argument since the load per pillar in those small churches was modest but above all because the proportion of facing stone in the pillar as a whole was much greater (Figure 62) than in the case of the Pantheon. It was an inappropriate application of the rules of proportion: the eighteenth century had seen spectacular progress in technical research but in a context without a genuinely scientific basis. Gauthey turned out to be the evil genius of Soufflot.

In 1771, a year after Patte's lampoon, Gauthey published a memoir on the application of the principles of mechanics to the construction of vaults and domes (*Mémoire sur l'application des principes de la mécanique à la construction des voûtes et des dômes*). Refuting the views of Patte, he went so far as to assert that four groups of three isolated pillars instead of pillars with integrated columns would

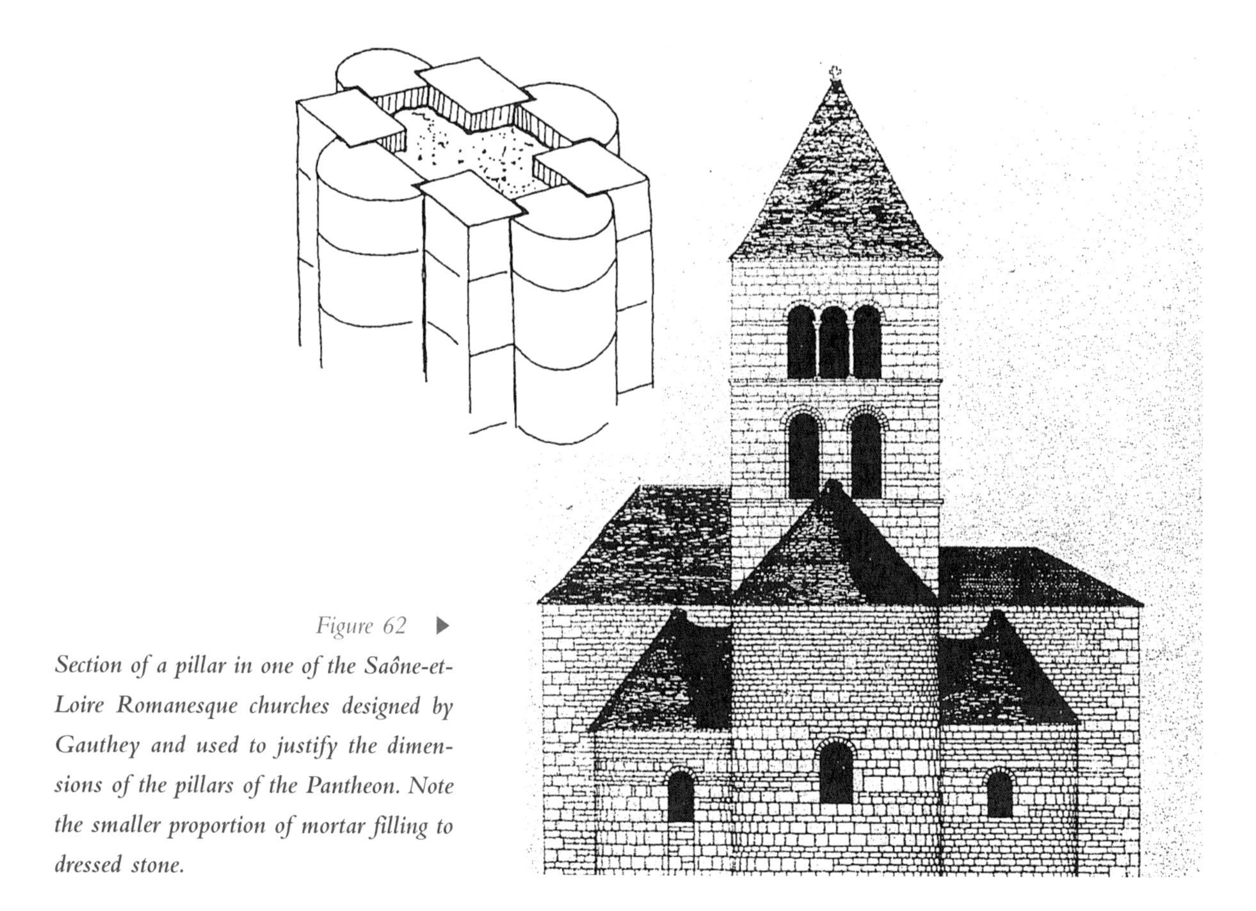

Figure 62 ▶
Section of a pillar in one of the Saône-et-Loire Romanesque churches designed by Gauthey and used to justify the dimensions of the pillars of the Pantheon. Note the smaller proportion of mortar filling to dressed stone.

suffice. Reassured, Soufflot hesitated no longer: wishing to avenge the taint on his honour, he made a wager of 24 000 livres with Patte, writing: "*I shall see to it that, within three or four years, he is covered with the humiliating shame I am preparing for him*".

Alas for Soufflot, the stones rapidly proved him wrong. Before even the first part of the dome had been completed, the first cracks appeared in the pillars. Utterly downcast, Soufflot died in 1780. The following year Patte triumphantly observed that 80 stones had split at the bottom; a later report noted up to 367 cracks in a single pillar. In 1785, before the completion of the dome, the pillars had already subsided by 45, 50, 60 and 54 mm respectively. By the year IV of the Republican Calendar (1795) they had subsided by a further 27, 18, 60 and 50 mm respectively. For those horrified by the table printed in the *Journal de Paris* on 19 Vendémiaire – the first month in the French Republican Calendar – of the year V there was no hope of the monument surviving for posterity: "*This proud cupola is about to collapse; it is going to crash down in a resounding tumult, crushing in its fall all the buildings around it...*".

Joint commissions of engineers and architects[42] then set to work in collaboration with the son[43] of Soufflot and Rondelet, one of the dead architect's assistants[44]. The latter, having been taken into the confidence of the workers, was aware of the shortcomings of his boss, to whom he nevertheless bowed. It is not really possible to sort out the exact responsibilities of each protagonist; on his deathbed in 1829, Rondelet asked his son to delay for fifty years publication of the memoir he had written on the pillars of the Pantheon. The document has been lost.

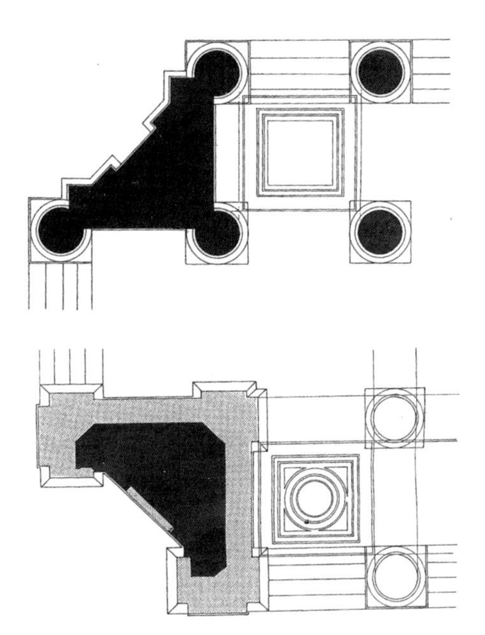

◀ *Figure 63*

One of the Pantheon's pillars before and after the restoration of 1809.

[42] Including Gauthey who had conducted some experiments on the resistance of stones to crushing and wanted to make up for his blunder of 1771.

[43] A mediocre engineer who was touchy and had a poor understanding of his father's faults.

[44] On this subject, see the rich archives of the Ecole Nationale des Ponts et Chaussées.

A forest of props was set in position and, after all the oak wedges had been taken out, an attempt was made to have the internal filling of the pillars do its share of the weight-bearing by sawing thin grooves around the circular perimeter of the facing stones. The core then compressed little by little and, as its section was considerable, it made an appreciable contribution to the overall effort. The building stabilized and no one realized that the top of the dome was now slightly lower than in Soufflot's original plans. Rondelet eventually thickened the pillars in order to hide forever the tell-tale signs of that disastrous period.

Other errors by Soufflot

Being an admirer of the Greeks, Soufflot inserted steel bars into the vulnerable parts of his building, even in the flat terraces, where the water stagnated and rusted the steel.

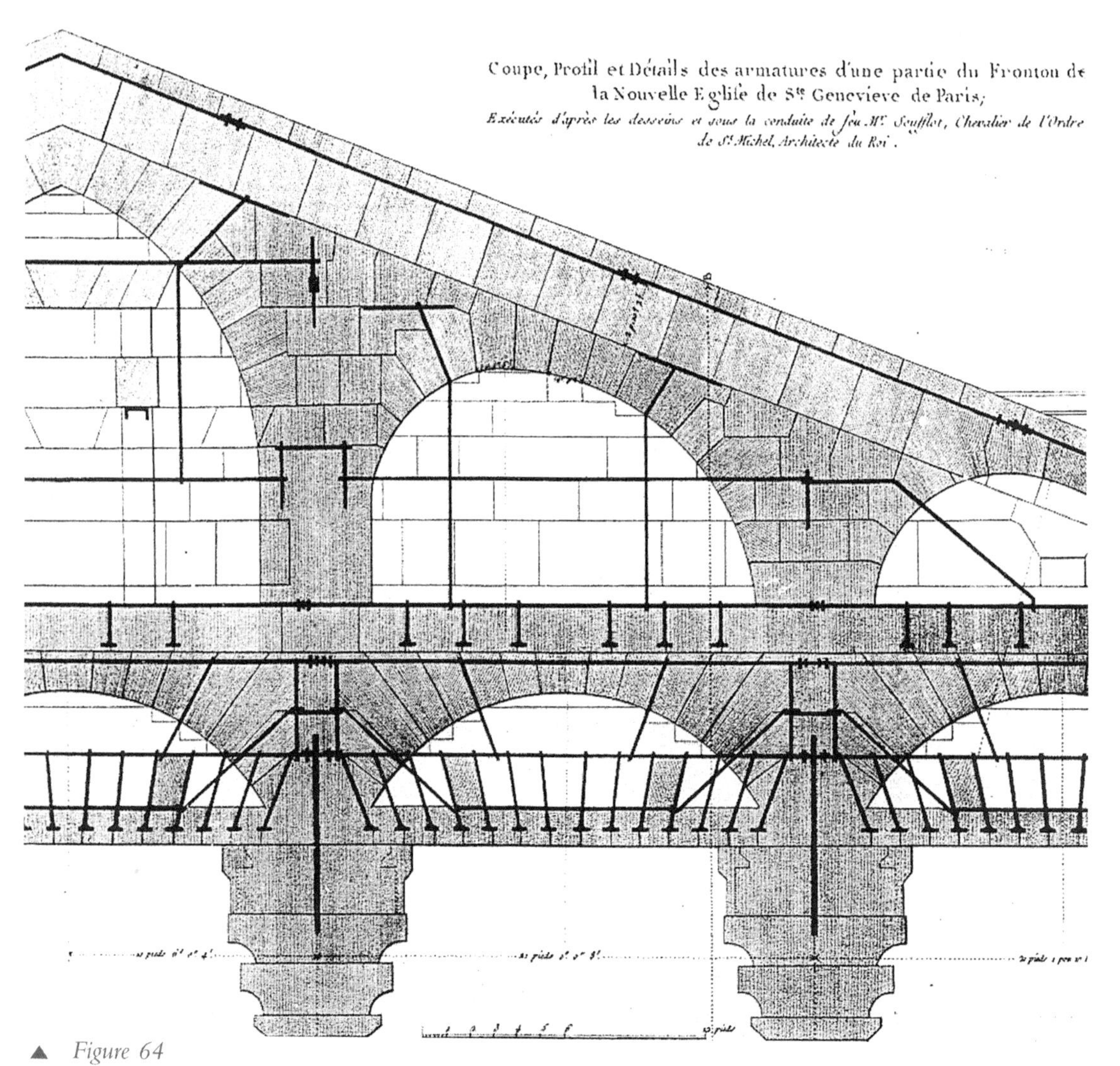

▲ *Figure 64*

Steel braces in the western pediment of the Pantheon (as published by J.B. Rondelet).

At the end of the last century the Pantheon found itself once again in peril[45].

As early as 1781 Soufflot had endeavoured to contain the thrust of the dome by having numerous windows in the nave and transept walled up. The Pantheon was shut in on itself, unable to see the street that bears the name of its luckless creator.

As can be seen from Figure 64, the pediments contain a large number of steel bars. No one would carp at their presence. But the rest of the building contains many other bars not indicated in the plans, especially under the vast flat open spaces of the terraces: their presence was occasionally signalled by the fall of a large chunk of stone that had become detached from the roof because of the pressure exerted by steel bars swollen by corrosion. One incident happened on the eve of a visit by the President of the Republic and sounded the knell as it were for the illustrious dead reposing for eternity in the crypt.

In Athens, Soufflot had admired the elegant entablatures of the temples but, being less prudent than the ancient Greeks, he had not enrobed the steel braces with lead; the internal parts of the building thus became vulnerable to water and condensation. On 2 December 1983, after four other similar incidents, a four-kilo block of stone crashed down on to the floor of the transept thirty metres below.

The dome has settled, numerous windows in the nave and transept have been walled up and the Pantheon has withdrawn into itself in its misery. The ashes of Soufflot were transferred there, but not until 1829, as if reluctantly; and yet his serious errors have not damaged his reputation and architects continue even today to laud his qualities. In 1970, at the Hotel Sully, with the collaboration and financial backing of the Caisse Nationale des Monuments Historique, a triumphant tribute was paid to his memory. The press went to town: "*Soufflot, the triumph of eclecticism .... The Pantheon of Soufflot is one of the high points, a summit in all senses of the word.... It epitomises not only French architecture but also the architecture of a large part of the world.... The Pantheon seems so grandiose and extraordinary that one can hardly believe in its physical existence*"[46]. This concert of praises echoes the words of the gallant Abbé Laugier in 1760: "*The monument will represent in the most distant centuries to come the foremost model of perfect architecture*".

O unmerited glory! Fortunately, all you stones, in whose fate the haughty Soufflot took so little interest, are still there to bear witness.

[45] See *Le Monde*, 5 and 6 March 1989, 'Le Panthéon en péril' by Jean Kerisel.
[46] From an article by Pierre Mazars in *Revue Arts*.

8

Stones and man in religious architecture: from moderation to excess

In Romanesque architecture all is wisdom and moderation and we feel that it has always been like that. The history of the three churches of the monastery of Cluny is highly instructive: two of them, Cluny I and Cluny II, were relatively unambitious. Cluny II was bigger than Cluny I but its nave was only 7.50 m wide and it had only one side aisle. The honour of building Cluny III was left to Saint Hugues. Known as the *Ecclesia Major*, it was a basilica that for six and a half centuries, until destroyed by the French Revolution, was the object of great admiration. With its 29.50 m from the crown of the vault to paved floor it was the highest of all Romanesque churches. Two side aisles bordered the nave and the church's full external width attained 39 m! These dimensions, which represented a satisfactory balance between reason and beauty, had been adopted by Abbot Hugues in 1085 only after forty years of travelling to all the capitals of western Europe. The *Ecclesia Major* was not finished until 1130 – long after the death of Hugues in 1109. His successor as abbot, the arrogant and impetuous Abbot Pons, was far from possessing the same qualities of wisdom and reflection.

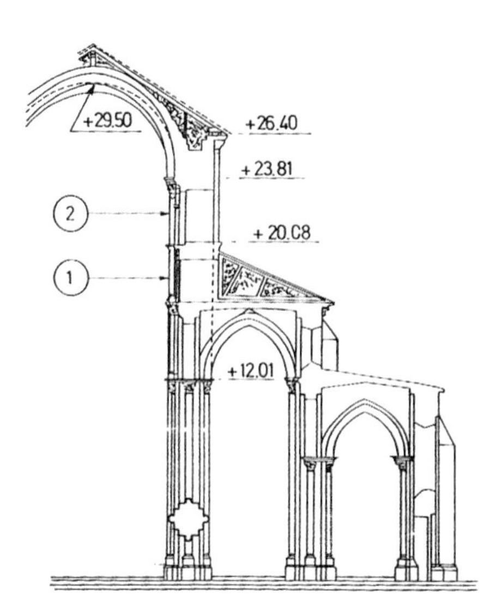

◀ *Figure 65*

Cluny III: Half cross-section of the nave, according to Conant, shortly before the accident of 1125.
1 and 2 draw attention to the slender solid stone columns in the triforium. Dimensions in metres.

Figure 65 shows a cross-section of the nave built by Abbot Pons: above the high lateral arcades there is a triforium surmounted by a high wall pierced with very large windows occupying no less than 48% of the surface of those lateral walls. For the very first time in the history of Romanesque architecture, the nave was flooded with light. It was a noble design but the walls, weakened by such large windows, were unable to support the powerful lateral thrust of the vaulted roof: in 1125 they gave way and several bays of the abbey-church collapsed. The disaster caused a great stir and the news of it even reached Rome. The Abbot who, with his usual modesty had quite unjustifiably proclaimed himself "*Abbot of abbots*", was condemned as responsible and transferred to Rome, where he finished his days in prison. In the building of his church, as in the pursuit of his career, he had cut too many corners.

The Venerable Peter, the last great Abbot of Cluny, rebuilt the church from the ruins left by his predecessor, who had omitted to provide flying buttresses ('1' in Figure 66) high up, indispensable for absorbing the thrust of the vault and transmitting it to the buttresses below.

The building work was completed in 1130. The result was magnificent. Recently, the general aspect of the abbey-church has been splendidly reconstituted by the American archaeologist K.J. Conant, who has devoted many years of study to the subject. To him, its destruction "*was one of the bitterest losses ever suffered by religious architecture*[47]".

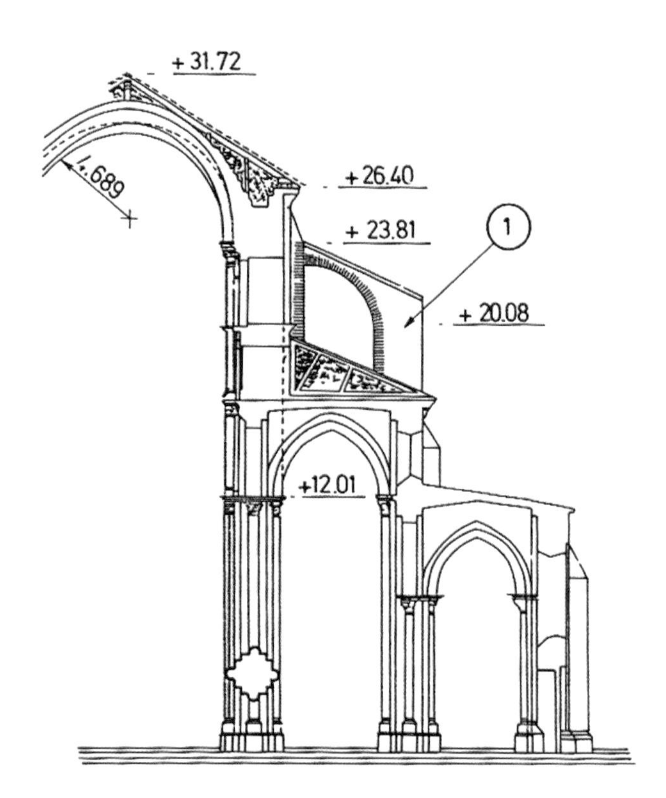

Figure 66 ▶

Cluny III: Half cross-section of the nave after the repairs carried out by the Venerable Peter in 1130. 1 = flying buttresses added to counter the thrust of the vault. Dimensions in meters.

[47] Conant, K.J., 1968: *Cluny, Les églises et la maison du chef d'ordre*, The Medieval Academy of America, n° 77, Cambridge, Massachusetts.

Figure 67 ▲

Photograph of the model of Cluny III, by the Caisse Nationale des Monuments Historiques.

Gothic

The lesson would not be forgotten: the accidents that had happened with Romanesque churches led to the high flying buttress and ribbed vault, which are fundamental elements of Gothic architecture, a new art that began at the abbey-church of Saint-Denis near Paris with Abbot Suger, a correspondent of the Venerable Peter. The stones were asked to make an extra effort, but it was a rational and acceptable one. In a single century, from 1170 to 1270, France would build eighty Gothic cathedrals and nearly five hundred churches. Each city wanted its own to be the highest and the most beautiful. All this did not proceed without mishap and, like the Romanesque style before it, Gothic architecture experienced a number of disasters caused by architects who had overestimated their own skill and the resistance of stones.

A case in point was the Cathedral of Saint Pierre in Beauvais. It owed too much to dreams of grandeur. The building was too spindly and ill-proportioned, both from the bottom up and cross-wise, for its ambition to be, at 48 m from keystone to floor, the highest of all Gothic cathedrals. It ended up as a half-church since it is still awaiting the construction of its nave.

The thrust of the vaults of its choir, which had been commenced in 1247, was originally contained by buttresses and pillars that were too widely spaced. One of them, perhaps two, collapsed

Figure 69 ▲

Saint Pierre de Beauvais: the counter-model of the pyramid. Steel bars and cables, open to the weather, come to the rescue of the lofty choir. (Photograph Emile Rousset, GEMOB.)

◀ *Figure 68*

The choir of Saint Pierre de Beauvais. (Photograph Emile Rousset, GEMOB.)

in 1284; the pillars were reinforced but the tops of these external supports were too flimsy: not so long ago they would still vibrate in stormy weather even though they had been interconnected by steel cables.

But funds were never able to match the ambitions of the chapter, which was reduced to building solely the choir. The transept was not constructed until two and a half centuries later and, to get it built, the people of Beauvais had to make pressing appeals to the Pope in an attempt to interest him in the highest monument in Christendom. Eventually, in 1518, Pope Leo X (1513-23) authorized the selling of indulgences and his two successors, Adrian VI and Clement VII, did the same. This outraged the monk Martin Luther and helped to trigger the Reformation.

In 1550, however, before the transept had been completed, the chapter had already decided to top it with a tower surmounted by a spire, a project that was particularly risky in the absence of a nave, which alone could counter the outward thrust of the choir: to build the tower before having constructed the first bays of the nave was a real blunder. One is above all struck not so much by the sense of grandeur of the Beauvaisians as by the feeling that they were simply trying to break records for height without giving proper thought to the matter.

It was very hard going. The chapter sought the advice of architects and carpenters before deciding on the materials to be used for the construction of the tower and the spire. In the end they adopted stone for the lantern-tower and wood for the spire, which would be surmounted by a large iron cross, giving an overall height of 153 m or just a few metres higher than the pyramidon topping the Great Pyramid of Cheops.

In 1571 it became necessary to take down the great iron cross, which was too heavy for the wooden roof structure : in high winds one could hear groans from the beams in the upper parts of the choir. The four pillars of the transept began to bend. In April 1572 they were respectively 5, 10, 15 and 30 centimetres out of true. The report drawn up by Gilles de Harlay and Nicolas Tiersault stated that "*the tower is beginning to lean for lack of buttresses and counter-buttresses*". On 18 April, "*the master-builder was sent to inspect the spire while the people assembled below. And as the clergy and the people were about to issue forth in procession, the master-builder, seeing that all was about to fall and realizing the imminent danger, and not having the time to descend, cried out in a mighty voice to those below to save themselves*".

Hardly had the clergy and congregation left the building than the spire, the tower, some of the pillars and part of the vault collapsed with a resounding crash and filled the church with "*a mountain of stones*".

Trembling in the wind

After the collapse of the lantern, the chapter of Beauvais remained in possession of an enormous structure that still rose high above the plain, though it had been seriously wounded by the weight and the collapse of that absurd tower. As the transept was not supported laterally by a nave, the pillars

Figure 70 ▶

The Cathedral of Saint Pierre in Beauvais as it was one morning in 1572. (Collection Bonnet-Laborderie.)

were forced considerably out of plumb, one of them by as much as 80cm, so that it leaned at roughly the same angle as the Tower of Pisa at its worst.

Whereas the current attempt to right the Tower of Pisa has been completely successful, the situation at Beauvais is, sad to say, deteriorating: precise measurements made by the National Geographical Institute show that, over the seven years from 1985 to 1992, the tops of nine of the thirty-six pillars that support the edifice have again shifted by several millimetres.

Figure 71 shows the poor layout of the support structures. You do not have to be an expert to see that something is not quite right in the bottom to top arrangement of the structure, which justifies the serious criticism levelled by Viollet-le-Duc in 1853: "*The buttresses intended to counter the thrust of the flying buttresses are abandoned to their fates for two thirds of their height.... The pillars of the choir, which are very long and slender and composed of thin courses of stonework, have necessarily settled under the weight of the vaults*". It goes without saying that there existed a profound contradiction between the long thin columns in solid stone, marked L in Figure 71, and the thick pillar A of stone encasing a rubble core.

In short, we find the same errors as found in the Tower of Pisa and the Pantheon.

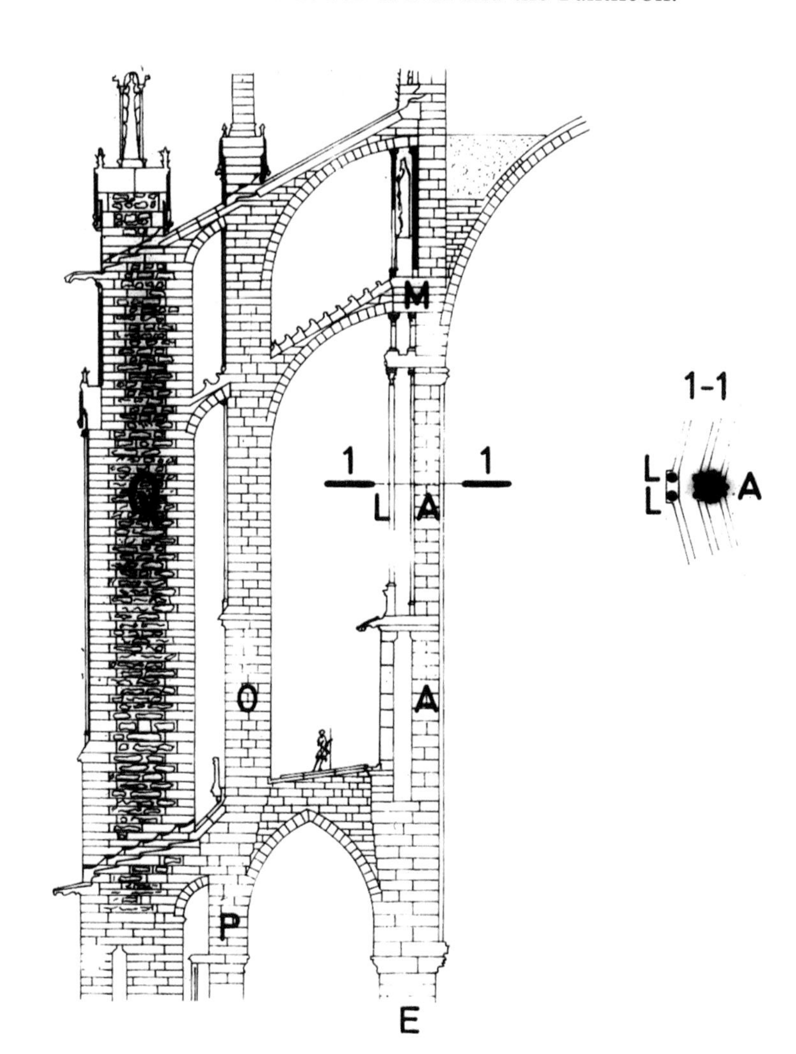

Figure 71 ▶
Section of the choir of the Beauvais Cathedral, drawn by Viollet-le-Duc in the nineteenth century, with a tiny human figure to suggest the scale.

Aerodynamics and architecture

But, even more serious, the external abutments, only 1.40 m thick for a height of 50 m, were laterally very unstable: they swayed in the wind. It became necessary to attach their tops to each other (done "*very badly*" in Viollet-le-Duc's opinion) by double clamps in iron in order to counteract the oscillation of the buttresses (see the horizontal projection of this clamping in Figure 71). A report at the time by a man called Baron and held in the Departmental Archives[48] is highly critical in this respect: "*I found it impossible not to censure the manner in which they were chained together to prevent the so-called oscillation in the wind; to obtain the desired result, this iron chain ought to have been inserted through the axis of the pillars at three quarters of their height.... The cost of this recourse to chaining, which may be estimated at approximately two thousand ecus, was very ill applied*".

A large part of this chaining[49] was dismantled in around 1960. Rightly or wrongly? It is an open question. And yet even today, in certain storms, the upper part of the transept is seen to sway and one can perceive cracks opening and then immediately closing up again. Sometimes currents of air flowing in certain directions provoke turbulence that resonates with the swaying of the abutments. There have been several serious alerts when the wind has been high and gusty.

In short, the structure is like a large disjointed sailing ship that groans and creaks in the wind, enveloped in a spidery filigree of unstable walls and steel cables of uncertain purpose, some taut as the strings of a guitar and others vibrating in the wind. It has proved impossible to correct the distortions in the masonry. Their persistence after so many centuries, falling stones and the loss of perpendicularity have produced a situation that has drawn leading specialists to the bedside of the sick building; their frequent disagreements show the difficulty of the problem and the widespread compassion for a building in such pain.

There is no doubt that, outside France, people tend to snigger at this unfinished and rather pretentious architecture and mock the sense of grandeur that often animates the French. At the University of Cambridge I once attended a lecture on the history of art and heard the lecturer poke fun at the great ship of Beauvais to the amusement of his students. Alas, the criticism was not entirely groundless, though it was hard for me to stomach it.

The most recent measures seem to be having a positive effect: instead of using flexible cables to tie together the external structures, it has been decided to traverse them with stainless steel tubes – a return to the chaining approach one might say, but more carefully thought out, as Baron would have wished.

Saint Pierre de Beauvais wanted, with its 48 m high choir, to surpass by far the Cathedral of Notre Dame in Paris (34 m) and four other great French cathedrals, Chartres (37 m), Bourges (37 m), Rheims (38 m) and Amiens (42.3 m). The spires of their cathedrals look down on so many centuries of history. Bereft of its lantern, Saint Pierre de Beauvais remains quite unique for its height, its lack of nave and its absence of harmony.

---

[48] Archives of the Oise Département, IV, 339.

[49] 'Chaining' implies the use of relatively inflexible bars of steel as distinct from cables.

What are the long-term prospects for Beauvais? I cannot help thinking of a trip I made to Dresden, where I had admired the celebrated painting by Bernardo Bellotto (1765) of the ruins of the *Kreuzkirche* of Dresden. This ancient Gothic church had been the target of a Prussian bombardment in 1760, during the Seven Years War. But the Venitian Bellotto, favourite painter of the Electors of Saxony, was perhaps partly inspired by the disaster at Beauvais two centuries before. I was already thinking of the subject that drives my thoughts today: the enslavement of the mineral kingdom, forced to submit to the ideas of human beings even when they are shown to be false. Stone has become the victim of extravagant dreams and castles in the air.

▲ *Figure 72*

Bernardo Bellotto: the ruins of the ancient Kreuzkirche of Dresden (1765). The author was struck by the comparison with a similar disaster at Beauvais two centuries before.

(Photograph Staatliche Kunstammlungen Dresden.)

9

Stones amid the waves

Let us now leave the vacillating heights of religious architecture and turn to the depths of the ocean. During the Renaissance, trade encouraged exchanges between Pisa, Genoa, Venice and Africa. Figure 73 reproduces a thirteenth century mosaic in the Basilica San Marco in Venice; as its Latin inscription indicates, it depicts the Evangelist St. Mark aboard a boat on its way to Alexandria. It was a voyage that excited admiration because it evoked the idea of Egypt: "*ad Egyptum per Alexandriam*" – to Egypt via Alexandria.

People would naturally speak of the Great Pyramid but also of another of the seven wonders of the world which Alexander and the early Ptolemies had decided to build on the Island of Pharos to make the approaches to that mysterious land safer. The Pharos of Alexandria was a lighthouse that gave its name in several European languages – 'phare' in French, 'faro' in Spanish, Portuguese and Italian, for example – to those lofty towers in or by the sea, some of them erected by men in peril of their lives in a semi-underwater world to guide navigators safely past hidden dangers.

In Book IV of the *Odyssey*, possibly composed in the eighth century BC, Menelaus describes the Island of Pharos to Telemachus in these terms : "*Now, there's an island out in the ocean's heavy surge, / well off the Egyptian coast – they call it Pharos - / far as a deep-sea ship can go in one day's sail...*"[50]. Mythology rejoined reality, though with a little poetic licence over the distance. One interpretation has it that one day's sailing corresponds roughly to the distance between the nearest mouth of the Nile (the Canopic branch) and the Island of Pharos. At any rate, the sediments of the Nile had certainly diminished the distance between the coast and the Island of Pharos substantially, so much so that it was decided to link the two by means of a mole some seven *stadia* – about 900 m – long.

The Roman poet Lucan, in his epic poem *Pharsalia*, thus relates the arrival of Caesar in Egypt: "*The zephyr stretched the sails and the seventh night revealed to him the flames of Pharos, lit up on the coast of Egypt; but the next day whitened the nocturnal torch before Caesar entered the calmness of the harbour*". The lighthouse of Alexandria still stood in the twelfth century, at the time of the Arab geographer Edrissi (1099-1164): its fire was called "*fanous*" and Edrissi wrote that to sailors the lantern appeared

[50] Homer, Odyssey, Book IV, lines 395-8. Translated by Robert Fagles.

like a star by night and a trail of smoke by day. However, we are told by Gaston Jondet, a French engineer in charge of Egypt's ports and lighthouses, that at the beginning of the twentieth century the entrance to the harbour of Alexandria was marked by no more than a simple beacon. That beacon, moreover, was overturned by the storm of 10 and 11 February 1911.

Where exactly was the famous lighthouse located? A memoir written in 1916 by that same Gaston Jondet, entitled "*Les ports submergés de l'ancienne île de Pharos*"[51] (The submerged harbours of the ancient Island of Pharos), summarizes all the historical sources and provides hydrographical data on the submerged structures and soundings. He shows the extent to which Pharos has been affected by historical events and modern building work. Alexandria and its harbours and canals are dependent on a subtle geology of local subsidence caused by the failure to drain the alluvial sediments and the dissolving of the local limestone through contact with air and water. This makes Napoleon's remarks on the wars of Julius Caesar more understandable: "*So, when all is said, there was nothing miraculous in all the war of Alexandria. All the plans drawn up by commentators to explain it are false*"[52].

▲ *Figure 73*

The voyage by St. Mark to Alexandria. (Photograph Scala Group.)

[51] Jondet, G., 1916: Memoirs presented to the Egyptian Institute, Vol. IX, Cairo.
[52] Napoleon I, 1836: *Précis des guerres de Jules César.*

In reality, the most recent underwater research at Alexandria has revealed that a rocky outcrop used to exist in the middle of the access channel, highly dangerous at the time since the level of the sea was then five metres lower and the surface of the land, and hence the top of the rock, about two metres higher than today. Indeed, it was quite possible that the lighthouse was built on that rock, thus dividing the wide entrance to the harbour, which stretched from the end of the heptastadion to the rocks of Cape Lochias, into two channels.

In Rome, a very different undertaking: the Ostia lighthouse

The Romans were faced with a difficult problem when the Emperor Claudius, in around 42 AD, ordered the building of a large port at Ostia. Paying no attention to the advice of Vitruvius never to build a port at the mouth of a river, he asked his architects to construct moles beyond the mouth of the Tiber with an opening accompanied by a lighthouse to show mariners the way to the great port of Rome.

According to Suetonius (70-ca.160 AD) in his "*Lives of the Caesars*" (Claudius, XX-3), "*At Ostia, Claudius threw out curved breakwaters on either side of the harbour and built a deep-water mole by its entrance. For the base of this mole he used the ship in which Caligula had transported a great obelisk from Heliopolis; it was first sunk, then secured with piles, and finally crowned with a very tall lighthouse – like the Pharos at Alexandria – that guided ships into the harbour at night by the beams of a lamp*"[53].

Recent excavations on dry land, as well as the description given by Pliny the Elder, make it possible to say that the flagship displaced 800 tonnes, was 104 m long and 20.30 m wide, and had six decks. After the decks had been dismantled, the ship was filled with ballast, scuppered and positioned on the sea-bottom at a depth of 6 m. The lighthouse was built on the prow of the ship (Figure 74). The open water between the lighthouse and the coast was then barred by sinking four smaller vessels to provide foundations for a new quay. The harbour and its lighthouse were long praised by Roman writers. Juvenal in particular relates the story of the return of his friend Catullus and the dangers he had faced during a storm: "*The storm subsided and hope was reborn with the sun. Then appeared the hills on which the city of Alba was founded. At last Catullus' ship passed the Tyrrhenian lighthouse and entered the harbour of Ostia whose defences, continuing beyond the lighthouse, keep out the waves of the sea and seem as if they want to flee from the shores of Italy, superb constructions more admirable than harbours hollowed out by nature*".

In my mind's eye, I can see that flagship, which preferred immersion in the silt of Ostia to ensure a safe passage for sea-goers to the joylessness of a marine graveyard.

Gradually, as trading by sea developed, the number of shipwrecks increased, with hulls stove in by unseen rocks hidden just below the surface. Towards the end of the seventeenth century an attempt was made to put an end to these disasters: the idea was to top these rocky outcrops with tall slender towers in fine stone called lighthouses, a revival of the ancient construction built to spare shipwrecks at the entrance to the Portus Magnus at Alexandria.

---

[53] Translated by Robert Graves.

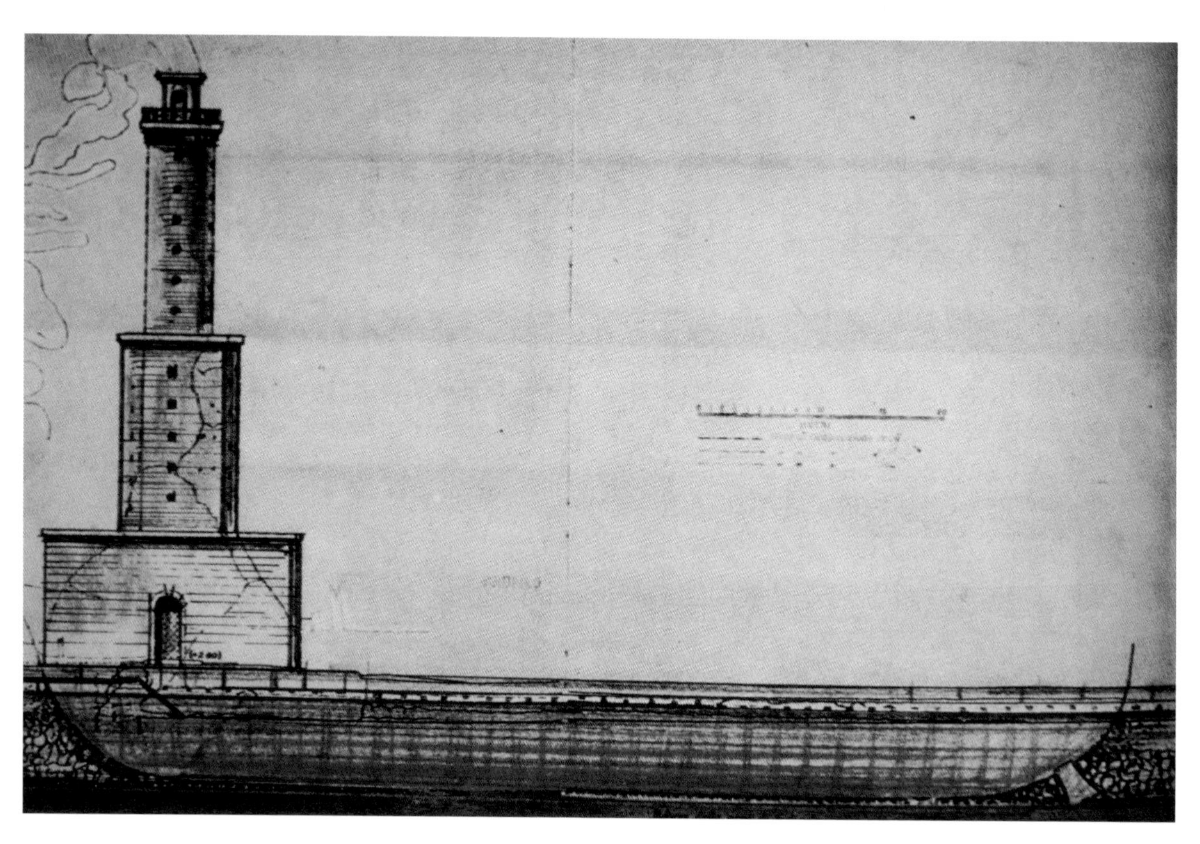

▲ *Figure 74*
The lighthouse of Ostia.

Figure 75 ▶
The first Eddystone lighthouse, 1696.

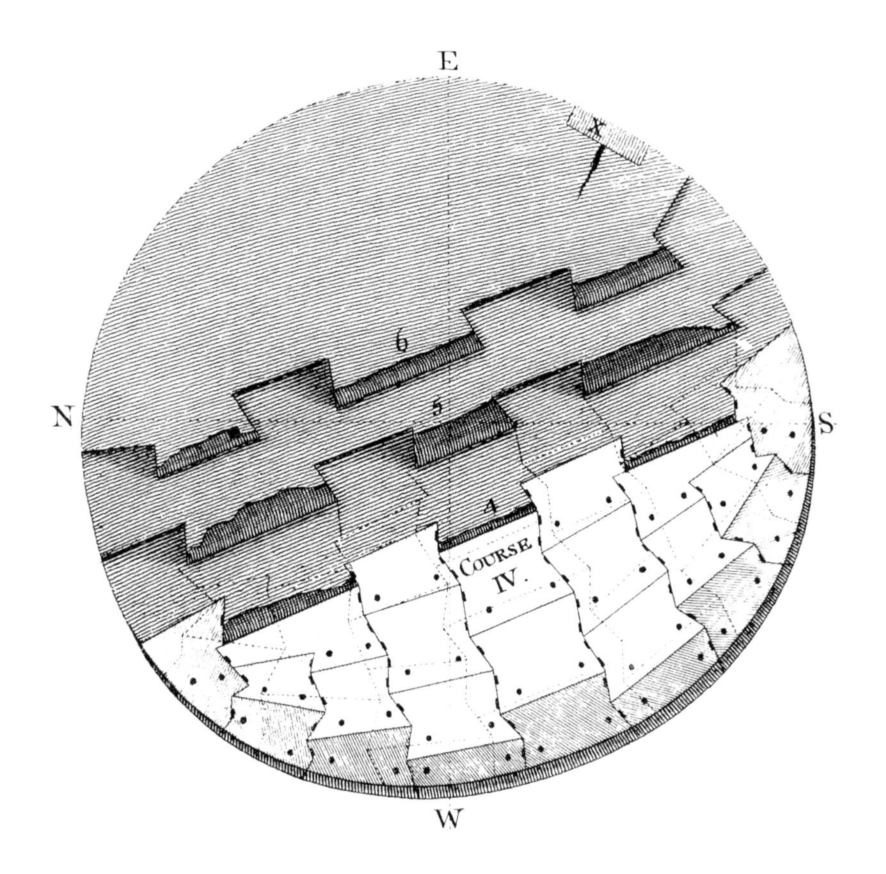

◂ *Figure 76*
Eddystone, foundation of the lighthouse: course No 4. The stones were cut with precision on the mainland and then fixed to the rock by dovetailing. The dots represent steel anchors to the rock.
(Royal Scottish Museum.)

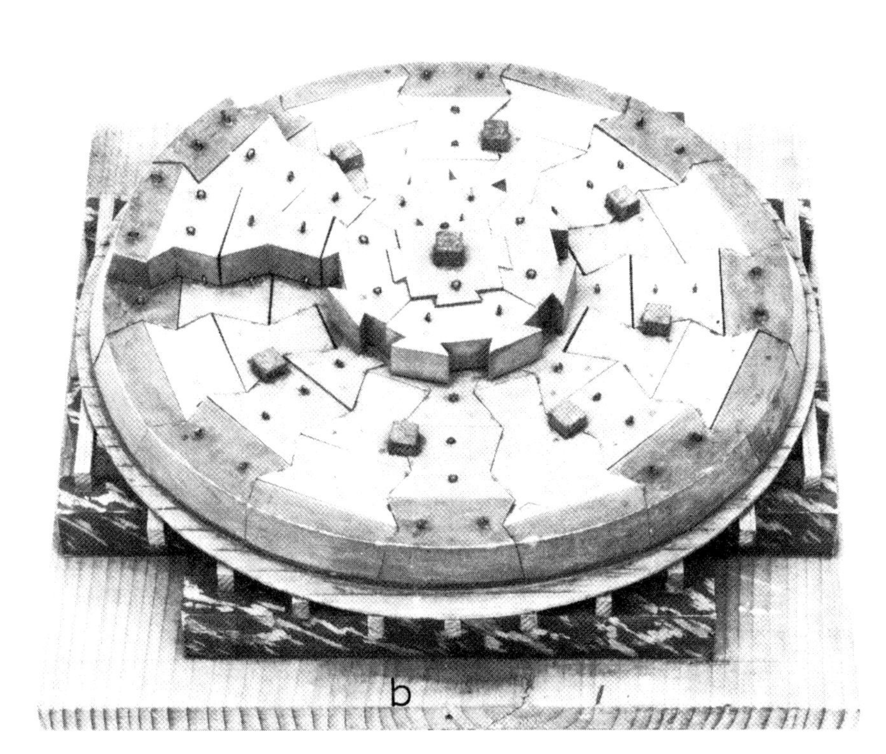

◂ *Figure 77*
Course No 7, not in immediate contact with the rock. Here the stones are also dovetailed. In addition, dovetailed joints connect two consecutive courses.
(Royal Scottish Museum.)

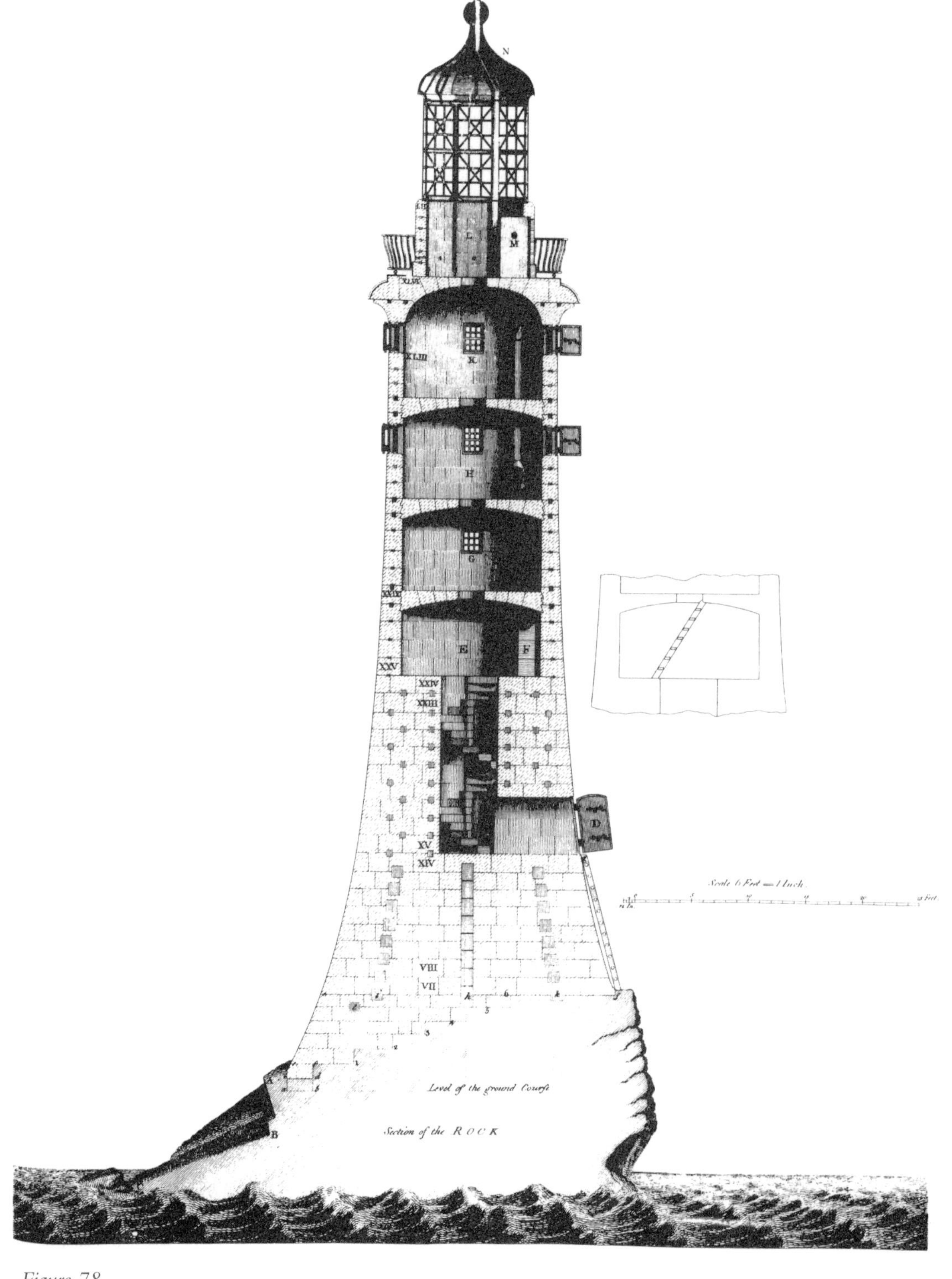

▲ *Figure 78*

Eddystone. East-west section of Smeaton's lighthouse in 1759, at low spring tide. (Engraving by E. Hooker from a drawing from Smeaton's "Narrative", plate 19.)

Thus, before the seventeenth century, ships setting their course for Plymouth had to find their way through the Eddystone Rocks some ten nautical miles south-south-west of the harbour. These storm-beaten rocks were completely covered at high tide. On the largest of the rocks, whose jagged shape broke the waves, the idea was born of building a lighthouse.

It was difficult to land on the rocks even at low tide. Four lighthouses in succession tried to gain a footing. On 20 November 1703, the sea swept away the first of them, along with the unfortunate author of the project, Henry Winstanley, his assistants and the keeper of the lighthouse.

For the second lighthouse, constructed by Rudyerd (1706-1709), the jagged tooth of rock was cut into steps on which a mixed structure of masonry and woodwork was erected and securely fastened to the rock with bars of steel sealed with lead; the covering of the lighthouse was of wood. Unfortunately, on 2 December 1755, the lantern set fire to it. The two keepers sought refuge on the external iron staircase while trying to dodge the falling drops of molten lead that dripped from the red-hot seals of the upper floors. The fire was so violent that it was seen from Plymouth. One of the keepers died ten days later and in his stomach were found seven ounces of lead; the other one, terrorized, escaped from the hospital and was never heard of again.

Not everyone was equipped to build a lighthouse. After these tragic events, John Smeaton, a remarkable engineer, found an intelligent answer: neither bars of steel nor lead are the friends of stone; stone itself must be used to bind stone to stone. He sculpted the rock by chiselling out the complementary shapes necessary to fix pre-cut stones by means of dovetailed joints, so that the stones would be securely anchored to the rock (see Figures 76 and 77). This system of dovetails in stone resulted in a perfectly monolithic base to confront the fury of the sea. The stones were solidly joined together and the result was so perfect that the lighthouse, completed in 1759, vibrated in unison with the waves that beat upon it. An indisputable victory of human intelligence.

But the upper part of the lighthouse was deliberately dismantled when in 1882 Sir James Douglas built the fourth and last Eddystone Lighthouse on a favourable rock a short distance away by constructing a cofferdam and pumping out the water, a method used in the far distant past, at the time of the Romans, by Vitruvius.

It was at virtually the same date that the lighthouse of Ar Men, whose fixing to the seabed also had many ups and downs, was inaugurated. West of Brittany, or more accurately just off the Pointe du Raz headland, lies an area of rock around the Ile de Sein and a few isolated snags called the Sein shoals, which vessels, sailing up from Gibraltar or the Cape of Good Hope and wishing to enter the English Channel, have to round by the west. It was most probably the perils of this area that gave its name to the bay on the mainland opposite – the Baie des Trépassés (Bay of the Dead).

One of the most seaward rocks that just breaks the surface at half-tide to the west of the shoals is Ar Men, 5½ nautical miles west of the Ile de Sein; on this rock, towards 1860, it was decided to build a lighthouse. But whereas the Eddystone rock emerged from the sea by 4.60 m at neap tide low water, the Ar Men rock emerged by no more than 60 cm and the current was still powerful. One immediately understands why the Bretons gave it the name Ar Men, which means "the Stone"; this simple title, contrasting with the more usual custom of qualifying rocks with adjectives, expresses an apprehensive dread of stones in the sea.

The construction began by the fixing of the foundations to the rock. In the first year, 1867, it was possible to land on the rock to work for only a total of eight hours. Progress was slow:

Year	1867	1868	1869	1870	1871	1872	1873	1874	1875
Landings	7	16	24	8	12	13	6	18	23
Hours worked	8	18	42	18	22	34	15	60	110

The inauguration took place in 1881 but the lighthouse caused concern on account of its diameter, which was only 7.20 m for a height of 34.50 m. In 1923, in the middle of a storm, Ar Men caught fire and the interior was destroyed.

It has been a curious combat fought against these underwater rocks, which prefer to rip the hulls of passing ships rather than allow themselves to be topped by towers.

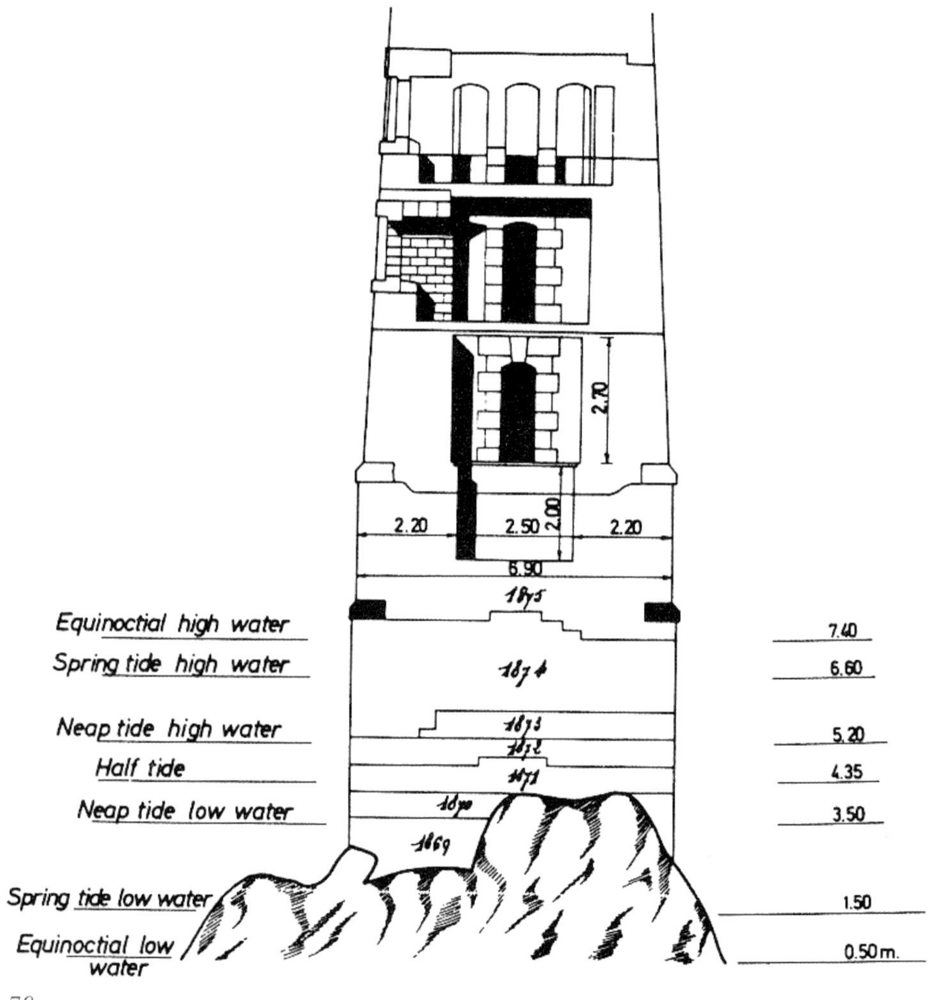

▲ *Figure 79*

Vertical section of the Ar Men lighthouse during its construction, based on a wash drawing of the period.

Figure 80 ▶

Photograph of the Ar Men lighthouse.

R-MEN

10

The Panama Canal and the soapstone of Culebra: a gigantic swindle

After he had successfully pierced the Suez isthmus, honours were heaped upon Ferdinand de Lesseps. It was just after their defeat in the battle of Sedan in the Franco-Prussian War and the French wanted something to admire. They called him "*the great Frenchman*". The Academy of Science welcomed him as a Member in 1873 even though he was far from being a scientist; the Académie Française coopted him in 1884 even though his style of writing was careless and not such as to win the approval of a literary coterie.

Other isthmuses attracted him: was he not known as the piercer of isthmuses? Mention was made of the isthmus of Corinth. But it was too short to interest him. Rather than reduce the time taken to sail round the Peloponnese he would join up two great oceans, the Atlantic and the Pacific.

Thus began the adventure of Panama. History has been excessively indulgent towards Lesseps, seeing him as the innocent and unconscious victim of unscrupulous financiers and the pocketers of graft money. The *Bibliothèque Historique des Editions Littéraires de France*, under the signature of Jean d'Elbée, has devoted a book to him entitled *Conquistador de Génie* (Conquistador of genius). His most recent biographer[54] goes so far as to cite him as an example for French youth and to speak of unjust judges at his trial. As André Siegfried has so rightly stressed, Panama is also the history of a canal and, to me, this history shows how arrogance and the will to excel can drive an old man to murky extremes in which intellectual honesty is sorely tried. The consequences were immense: it cost about ten thousand lives, and the eventual loss, amounting to a billion and a half francs, that is three times the cost of the Suez Canal, ruined the personal savings of the French.

At Suez, Lesseps, surrounded by brilliant engineers, had given a false impression. He had very mixed views concerning his engineers, considering that they took delight in complicating matters; he considered it better to trust in his own instinct, in his common sense and in his talent for negotiation. At Suez, a few well-placed pawns, charming smiles, promises of "*sweeteners*" and letters to the Empress were to his mind just as important as the researches of Lavalley, the engineering genius who

[54] Ghislain de Diesbach, *Ferdinand de Lesseps*, Paris, Perrin, 1998, pp. 17, 403-23.

invented the dredging machine with its lateral discharge, without which the canal could not have been dug.

But the Empress disappeared in the political whirlwind of 1870: in Panama, Lesseps had to fall back on his reputation as "*the great Frenchman*". So he reasoned by analogy – a gigantic risk when dealing with a project dominated by subjects such as climatology, epidemic medicine, hydraulics and the earth sciences.

He brushed aside all objections with a "*I did Suez and I know what it is about*". He would not listen to anybody; he had decided that the canal in Panama, like its twin at Suez, would be at sea level, without locks. If it was objected that there was a difference of six metres between the spring tides of the Pacific and Atlantic, he replied that the geographers must have made a mistake, as Lepère had done at Suez, and that locks were as pointless as the ones imagined for Suez by the Saint-Simonian Talabot. If it was objected that there would be a lot of earth-moving to do to dig through the 100 m high Culebra ridge, a formidable challenge when the rock concerned became like soap in the rainy season, he would reply that he had got his way with the 17.50 m Gisr ridge that had lain in his path at Suez. He was told that a river, the Rio Chagrès, which could become a raging torrent, crossed the route of the future canal; there was nothing like it at Suez. He was told that the rock at Culebra became like soft paste in the torrential rains. Did people really believe that he knew nothing about the tricks of the soil, when he had quickly vanquished an unexpected bed of rock? He was told that the murderous climate had already engulfed thousands of human lives in the building of an inter-ocean railway in Panama. Did people really believe that the digging of the Suez canal under a blazing sun had been a party of pleasure?

In vain was it explained to him that one must adapt to nature, explore it thoroughly and not do violence to it. In vain, too, did Godin de Lépinay, a great engineer who had worked in the isthmus of Panama, tell him of the very unhealthy climate and point out the shortcomings of his project. Godin de Lépinay even developed a simple and rational counter-project using locks which the Americans[55] would later adopt when they took over at the beginning of the twentieth century. The celebrated Gustave Eiffel associated himself with all these criticisms during the discussions of an *International Commission* that met in Paris. One hundred and seventy-five of its members, mostly from abroad, unconditional admirers of Lesseps and carefully selected by him, who had never set foot in Central America, overrode the criticisms of Godin de Lépinay and Eiffel. Before the vote, Godin de Lépinay made a solemn declaration: "*Not wishing to burden my conscience with all these deaths to no purpose and with the loss of a considerable outlay... I shall vote no*". A prophetic warning but without effect owing to the pressures exerted by the "*great Frenchman*".

At this point Lesseps became guilty of serious misinformation that almost amounted to a confidence trick. He sent a few samples of hard stones (probably from the ballast of the inter-ocean railway) to the Academy of Science, stones that were but remotely connected with those of Culebra.

[55] It is quite wrongly asserted that the Americans took up the project of Lesseps. They had always considered it unrealistic. Bunau Varilla, whom they called the 'Bonaparte of engineers', saved French honour by persuading them to adopt, some twenty years later, the project of Godin de Lépinay.

The illustrious Academy[56] then certified that the deep cut at Culebra could be realized by their fellow Academician with no particular difficulty and with almost vertical sides. I found this procedure so dishonest - even criminal – that I went to the Library of the Academy to check the matter. At that time the natural scientists outnumbered the other Academicians and knew precisely nothing about the difference between the volcanic but saponifiable rocks of the Culebra Cut and the particularly hard rocks of the Corinthian Canal. The Academy made a very serious mistake and there is even a suspicion that it was once again our great Frenchman who, as with the International Commission, had wielded the pen!

It should be noted that nowadays the learned assembly usually refuses to give an opinion on matters of this kind. One can sense the insistence of Lesseps when one reads the astonishing opinion it emitted at the time: "*The design and workplans of this enterprise are worthy preparations for an undertaking that will benefit the whole of mankind*".

Lesseps would later address the Grandes Ecoles, which had provided so many excellent engineers for the Suez project, telling their students what an exalting task awaited the volunteers and how healthy was the theatre of their future exploits (Lesseps himself paid three visits to Panama, but each time in the dry season when there was little yellow fever).

So many executives and workers died there that the Americans called Lesseps "*The great undertaker*"[57]. "*We arrive together*", an engineer named Petit remarked to a young graduate of the Polytechnic called Sordoillet, "*and we shall die together*". And that is what happened. Yellow fever was raging: every morning a train would carry the bodies of those who had died in the night to the cemetery of Colon. The deeper they dug the Culebra Cut, the more landslides there were and the more material was left to be excavated. According to Lesseps, the Cut should be stable with almost vertical sides. In the event, they were softened by the rains and springs so that, in the end, the slope was reduced to only one in four. The French began work in 1882 and gave up in 1889. In spite of all the disasters and even when everything was going wrong at the site, the communiqués to shareholders and future subscribers were always optimistic. In 1888 Charles de Lesseps, the son of Ferdinand, declared at a banquet for contractors: "*The Canal will be opened in 1890. We have altered our work programme. Since the mountain would not come to us we have gone to the mountain.... I drink to the health of the French workers*".

The French gave up after having excavated only 14.5 million cubic metres, to be compared with the final total of 73.5 million by the Americans for a canal that was not at sea level and hence less deep.

Lesseps, an old man basking in his glory and exceedingly pig-headed, totally lacked moderation and critical sense. Speaking of *hubris*, Aeschylus had this to say: "*It is an evil familiar to everyone from having cursed it or committed it. It is a thought of pride resting on idolatry or on the cult of one's own will. Not to be drawn into it is the only way of escaping Nemesis, the divine vengeance*". Lesseps would not escape

[56] Minutes of the Academy, 1880, 2nd semester, pp. 272, 274-5.
[57] See McCullough, D., *The Path between the Seas; the Creation of the Panama Canal (1870-1914)*, New York, Simon and Schuster. McCullough has written an excellent biography of Lesseps during the second half of his life.

the ire of those despoiled of their savings. Already much weaker physically, he was brought to trial and initially condemned to five years of imprisonment, as was his son Charles, who had warned his father from the outset. Gustave Eiffel had, as a last resort, been caught up in the wake of Lesseps and persuaded to design the locks to which the latter, his back to the wall, had finally resigned himself; he had pocketed a considerable sum of money on account and was sentenced to two years. "*Beautiful delusions*" conceded the lawyer defending Lesseps to the Court.

France was impoverished and Egypt ruined. The dying Lesseps had lost his honour and tarnished all the glory he had gained in Egypt, while a steadily growing traffic through the Suez Canal would prove that he had been right after all, the magnificent promoter of an idea, to trace that blue ribbon dreamed of by the Saint-Simonians over the soil of Egypt. As for the worksite in Panama, it quickly disappeared into the jungle like a lost civilization, carrying with it the illusions – and the savings – of so many ordinary people and nipping in the bud the careers of so many promising young men. And yet, for many people, Lesseps remains the great Frenchman to serve as a model for youth. The deserted Culebra and the cemeteries of Colon with their thousands of graves remain silent.

In speaking of this sorry episode, I recall a few pages (from page 172 onwards) in the book by David McCullough *The Path between the Seas: the Creation of the Panama Canal (1870-1914)* devoted to life at the Ancon Hospital at Colon and the Sisters of Charity, only two of whom survived. One of those two was the Mother Superior, Sister Marie Rouleau, who in 1868, when hardly adult, was sent to Panama by the Hospital of Versailles several years before the first engineers arrived. She was a woman of exceptional courage, known and admired by everyone. What about you, Mr Lesseps?

11

Works of stone engulfed by sand or water

Beneath the hum of the world is the silence of cities engulfed by sand or water.

The sand carried by the winds that sweep across the desert covers with an ever-thickening shroud thousands of square kilometres of land. In China, it devours some 2500 km^2 per year and already covers 27.3% of the country; a giant dune is threatening to bury Beijing in dust. A desert is on the move.

Often, however, time works differently in its efforts to efface the traces of man. The water level may rise, the land sink or glaciers melt. Slowly but surely the marks left by man are erased, at a pace barely perceptible to the human eye. "*I have seen dry land become a sea*", wrote Ovid (43 BC–18 AD). I myself was first made aware of such changes wrought by nature by Pierre Termier, an eminent professor of geology at the Ecole Nationale des Mines in Paris, who used to describe the slow but inexorable invasion of the ice age glaciers as they moved south from England towards France and the future Lutecia (now Paris). He had such powers of evocation that his students, of whom I was one, shivered with the advancing cold and pulled their clothes more tightly around them as he brought his lecture to an end.

Engulfed civilizations are legion around the world; I shall deal only with a few on the shores of that sea which the Romans called *mare nostrum*. Only ten thousand or so years ago the Mediterranean was 100 m lower and, to judge from the relief of the sea-floor, there were in fact two seas for what is now Sicily was almost welded to Tunisia. The entrance to the Cosquer Cave (Figure 81), named after the diver who discovered it near Marseilles in 1991 and now 36 m below the present sea level, was then well above the water line. A rising gallery some 175 metres long leads to an immense cavity, the upper part of which is still above sea level and has walls covered with the negative impressions of hands and many realistic images. The main periods of artistic activity were around 27 000 BP and around 19 000 BP, at which dates the sea was respectively 70 m and 100 m below the present level. It is difficult to see the logic in this. Perhaps in 27 000 BP, the rising water threatened the entrance to the cave, which those human beings then began to decorate. Subsequently, from 19 000 BP, the rise was rapid before slowing down more recently.

Far away to the south-east, a canyon with almost vertical sides had eaten into the Eocene lime-

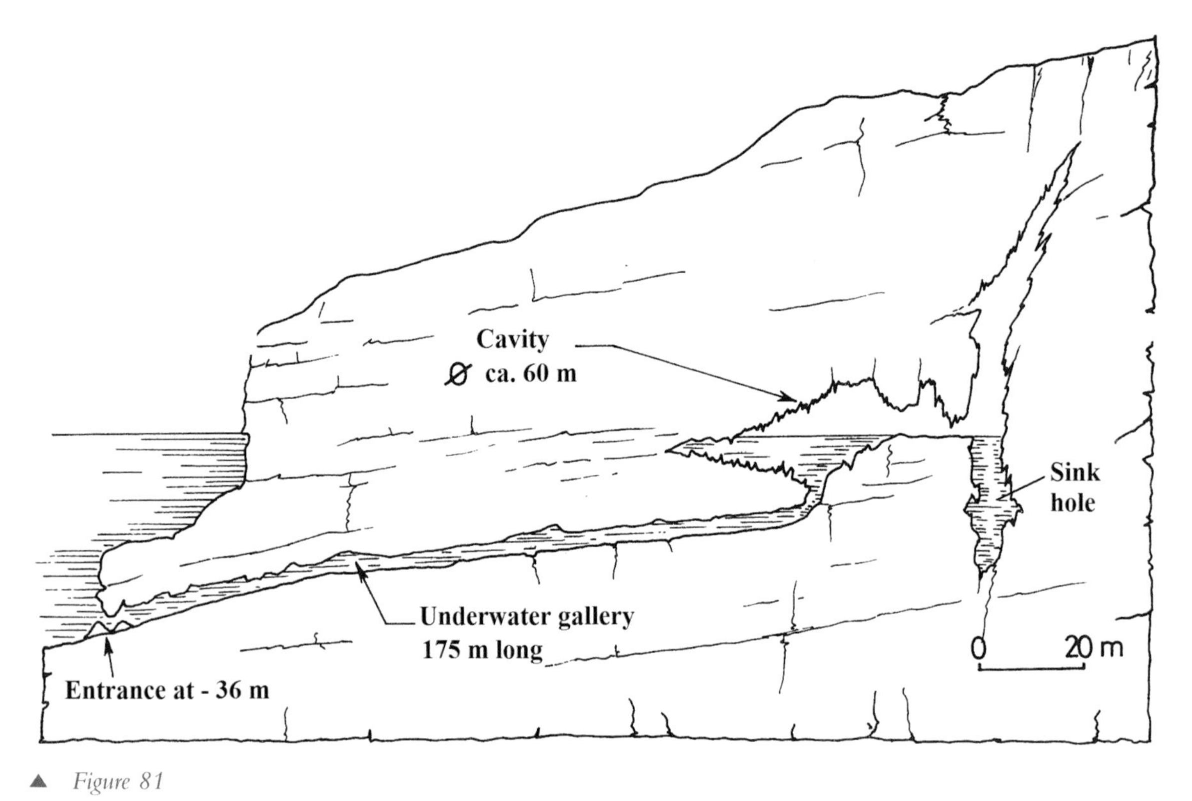

▲ *Figure 81*

The Cosquer Cave, near Cassis, in the vicinity of Marseilles, France.

stone at the mouth of the largest affluent of the *mare nostrum*, the Nile. In those cliffs along its banks lived troglodytes. The sediments carried by the seasonal floods of the Nile gradually filled in the canyon and created the first areas of cultivable land, which were later drowned once again as the level of the sea rose.

This variation in the level of the Mediterranean is confirmed by the remains of numerous civilizations that were once dotted around its shores but have now disappeared under water. On the southern shore, for instance, lay the port of Appolonia (now called Sousse) in Cyrenaica. The ancient port included an outer harbour with access via a narrow channel between two towers to an inner harbour with quays and berths for ships. The coastline appears to have receded southwards about a hundred metres, causing the ancient harbour to disappear under 2-3 metres of water, with the ancient storage silos cut into the rock at the foot of the Acropolis now invaded by the sea.

Let us now turn to the northern shores of the sea to discover or, more precisely, rediscover the Roman fishponds dating from the first century of our era. In 1981, G.G. Schmied studied eight of the twenty-three found along the Tyrrhenian coast. One of them is the so-called fishpond of Lucullus, circular in shape and divided into four compartments for four different species of fish. Channels were cut into the rock and covered with vaults to provide cool areas where the fish could shelter from the summer heat. Other compartments contained amphora in which the fish would spawn. All these constructions are now under water, but not deep.

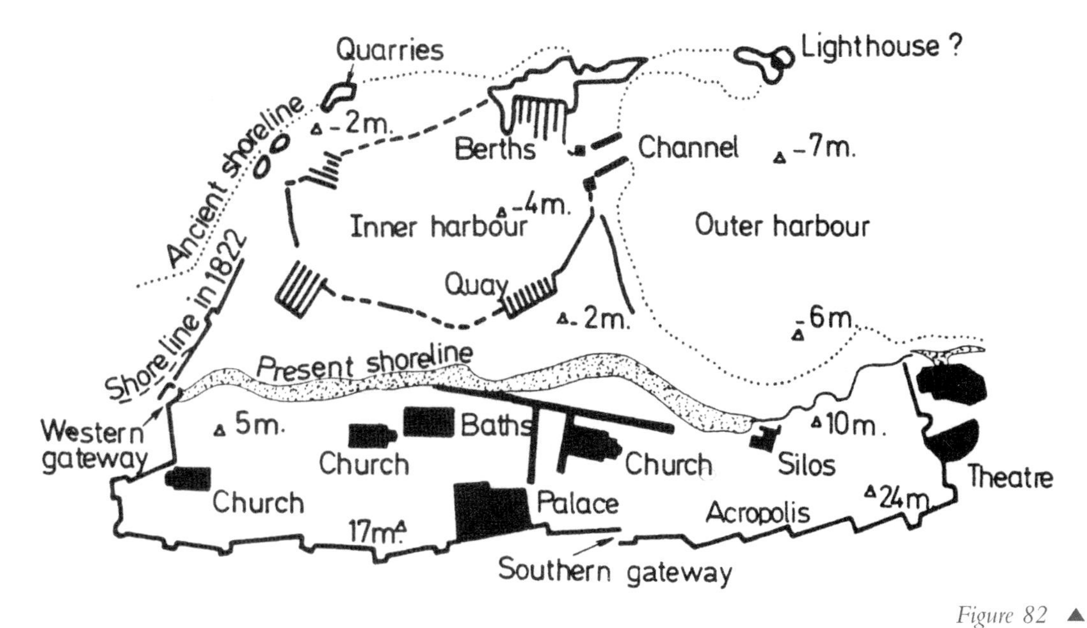

Figure 82 ▲

The ancient port of Appolonia in Cyrenaica, now submerged (Prof. André Laronde, 1981).

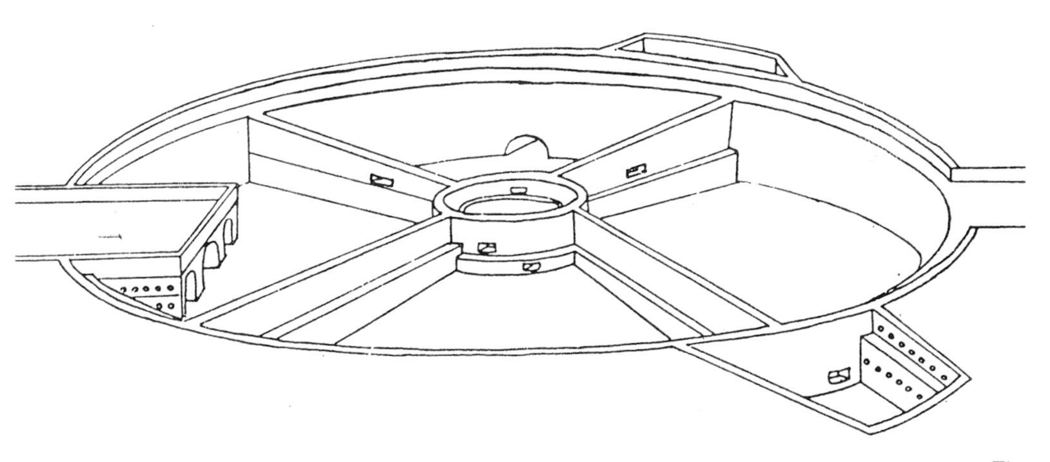

Figure 83 ▲

The fishpond of Lucullus on the Tyrrhenian coast. Oblique view from the east, after Chiapella.

The special case of Venice: a submersion largely provoked by human beings

The progress in underwater exploration has increasingly focused attention on drowned cities, cities once humming with life but now silenced for ever. In spite of such progress, however, it is very difficult to see them clearly: at Alexandria, for instance, the underwater remains of buildings and sculptures are in great disorder, jumbling the superposed remnants of successive civilizations with borrowings from previous periods in a chaos created by earthquakes.

On the other hand, the causes of the submersion of Venice are apparent in details that point clearly to the guilty action of man. The seabed in the Gulf of Venice is bordered to the west by a lagoon with three openings, the channels of Chioggia, Malamocco and the Lido (Figure 84).

At the time of the Doges, the lagoon used these openings to maintain a daily equilibrium through the to-and-fro movement of the tides. This equilibrium is now seriously threatened by the rising level of our *mare nostrum* but also and above all by recent constructions added for the development of tourism (road bridge in 1932 and railway bridge in 1946) and for the industrial development of Italy (Marghera, second largest port in Italy). The lagoon has now been gutted by a number of "*aquatic motorways*", dredged channels some fifteen metres deep, along which sail oil-tankers and cargo-

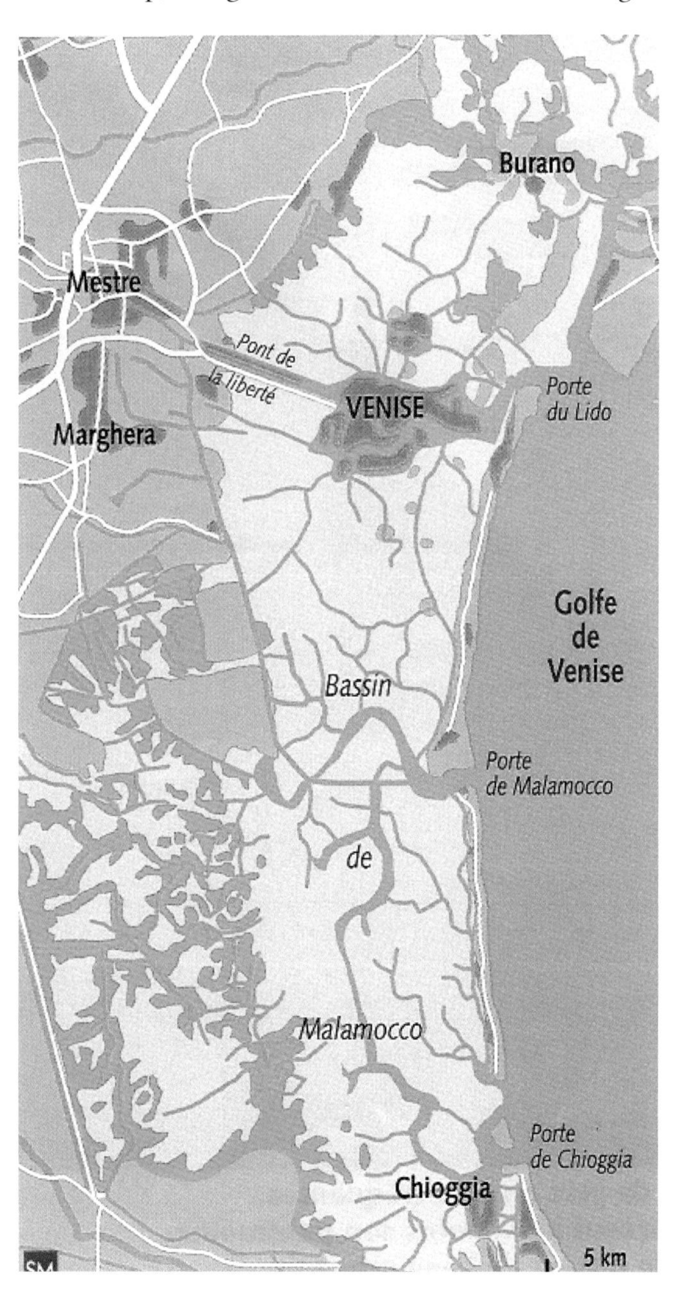

Figure 84 ▶
The three entrances to the Lagoon of Venice.

ships and into which pour enormous quantities of water.

Nature reminded man of the risks he was taking during the night of 3 November 1966. During a violent storm, giant waves stirred up the Adriatic and, for twenty-four hours, submerged the Piazza San Marco under 1.94 m of water, on the surface of which floated tons of refuse mingled with the dead bodies of animals. The electricity system broke down. The city of the Doges suddenly woke up to its vulnerability and its inhabitants started to leave. Out of a population of 175 000 in 1951 only 67 000 remain today.

Tourists do not always realize the gravity of the long-term situation; for them, the "*acque alte*" are as much a part of their stay as the trip up the Grand Canal in a *vaporetto*, but the alternate rising and receding of the floodwater is insidiously rotting the foundations of wooden piles on which the palaces we so admire rest.

The ecologists want the massive movements of water to be stopped by reducing the dredging of the lagoon to the minimum necessary for its health, whereas the Municipality and the State have opted for an eye-catching project, called Mose (acronym for *Modulo Sperimentale Electromecanico*), in which huge sluice-gates will close each of the three openings at each major high tide. It is a scheme that has been talked about for the last ten years and will cost around three billion euros. It is scheduled for completion in 2011, but there is no guarantee that it will put an end to the *acque alte*.

Tenacious legends and facts of history

Veneta, the Venice of the Baltic, which was swallowed by the sea in the twelfth century, was one of the biggest cities in Europe; a German archaeologist has announced its rediscovery on the banks of the Oder, in Mecklembourg-Pomerania. The legend of this city had inspired Selma Lagerlöf for her novel *The Wonderful Adventures of Nils*: a young shepherd, who heard the church bells of Veneta tinkle under the waves, sees for an instant the magic city gleam in the sun.

The city of Ys is the subject of another wonderful legend profoundly rooted in the imagination of old Breton-speakers : none of them would ever cast doubt on its existence or its beauty – as they say in Breton "*Abaoue eo beunzet Is, neus ker ebet evel Paris*" (*since the city of Ys was drowned, there has been no other city to compare with it*)[58]. Ys had a tragic fate; it was submerged to atone for the terrible misdeeds of Dahut, the daughter of King Gradlon, who reigned over Ys. She had been bewitched by the devil, who was known as Paulic.

The city was situated on the coast of Cornouaille, near Douarnenez. Its ramparts were of a wild beauty, being faced with the blue-grey granite found all over Brittany and the pink granite of Erquilanum, near Cap Fréhel. Disguised as a prince, Paulic succeeded in seducing Dahut and persuaded her, as a token of her love for him, to steal the golden key that King Gradlon her father wore around his neck while he slept. Dahut gave the golden key to Paulic who went to open the sluices that protected the city against the sea. Meanwhile the princess revelled in the palace with her boon companions. But Saint Guénolé, at full gallop on his horse, managed to pull up the old King Gradlon

[58] Déguignet, J-M., 1992: *Mémoires d'un paysan bas-breton*, An Here, 1992, p. 92. For Breton-speakers, there is a pun: 'Paris' means 'like Ys'.

behind him on to the rump of his horse and save him from the rising water. At the top of the Cathedral of Quimper, between the two towers, one can still see Gradlon on his horse, his eyes riveted on the sea as if he were about to see his city of Ys emerge from the waves.

Ernest Renan, in his *Souvenirs d'Enfance et de Jeunesse*[59], published in 1883, told this story, which contained elements that reminded him of his own life: "*One of the most widely known legends in Brittany is the story of the city called Ys which, at an unknown time in the distant past, is said to have been swallowed up by the sea. Various spots on the coast have been identified as the site of this fabulous city and fishermen tell strange tales about it. On days when there is a violent storm, they say, you can see in the hollows of the waves the tops of the spires of its churches; when the weather is calm, you can hear rising from the deep the sound of its bells, accompanying the hymn of the day. It often seems to me that deep in my heart there lives a city of Ys that still rings its obstinate bells calling to holy office the faithful who no longer hear*".

King Gradlon travels all the roads of Cornouaille in folksongs and storybooks. The city of Ys thus serves as the emblem of all engulfed cities.

Memphis

Legends are durable. History, on the other hand, sometimes passes by some of the finest moments in the life of the human race, moments of such peril that they become legendary. Here I am thinking of the inhabitants of the great Egyptian capital Memphis, which was for centuries the biggest capital city in the world. In the words of the Egyptologist Gaston Maspero, "*Memphis was for Greeks of the time of Herodotus what Cairo has been for those of today, the quintessential oriental city and eminent archetype of old Egypt. Despite the disasters that had befallen it in the preceding centuries, it was still very beautiful and, with Babylon, the largest city in the world*".

It had been built by 2400 BC, in the middle of the Nile valley, oddly enough at a point where the river narrows before spreading out into the delta. It was not a site to be recommended since the level of the Nile in flood rose considerably owing to the existence of this bottleneck, which got worse as the capital grew in size. Hydrologists and archaeologists haughtily ignore each other, the latter being quite insensitive to the danger that threatened this astonishing city every year and uninterested in the mighty walls that protected it.

There is every reason to believe that a stone masonry wall with a trapezoidal section, as high as six or seven tall men on top of each other, protected hundreds of thousands of human beings for nearly three millennia. At their peak, the floodwaters of the Nile would lap against the rampart walk on top of the wall. The population would then worry that the water, swollen by its passage through the bottleneck, might continue to rise. They would listen to the stones groaning under the pressure of the flood; they would see tell-tale damp patches as water seeped through the joints in the masonry and formed pools at the foot of the wall.

Bold boatmen sailing down the Nile, carried along by the powerful current, would hear the hum of the noisy city before they saw it at the bottom of an enormous stone bowl, with its teeming population trusting in the protection of their god Ptah, to whose stone temple they directed all their prayers.

[59] *Oeuvres complètes*, Paris, Calmann-Lévy, 1883.

◄ *Figure 85*
A drawing of naval games organized in the Rome Colisseum, when "the arena became a lake for naval combats". Memphis could well have looked a bit like this.

The divinity Ptah had an immense reputation. He was regarded as the creative Word, he who raises the firmament with his hand, and was addressed with the words "*Praise be to thee, Ptah, who contents the vital needs of men and gods*". And yet the inhabitants of the city could not have forgotten the great misfortunes that had befallen the city when the river in flood rose higher than ever before and traversed the parapet, drowning large numbers of people.

But such tragedies did not sap the energy of the population: the city was the Pivot of the Two Lands, and although it impeded the flow of the sacred river, no other site was better placed to symbolize the good relations between Lower and Upper Egypt. After each disaster, the pharaoh, the Word made Law in much the same way as Ptah, decided that an even higher wall must be built at all costs to encircle an even larger city.

As time went by, the wall became the wall of solidarity and perfection. The artisans who constructed it knew that the tiniest fault in the perimeter (some 11 000 metres in length) would endanger the entire city. Seen from without, the wall was a most impressive sight: to Pianki the city appeared impossible to capture when, towards 720 BC, he advanced from Upper Nubia to overthrow the pharaoh. On the stela that records his campaign can be read the words "*Memphis is fortified with a wall, a great rampart constructed with much cunning; it was impossible to attack it*". This important testimony shows clearly that the city had not been built on imported ballast like many of the farms in the valley.

The books on Egyptology say nothing about the succession of disasters that befell the city and the heroic life of its inhabitants: as if the discovery of a mummy or of a deposit containing a few statuettes made it pointless to try to ascertain the shape of the capital. The discovery of the wheel and the testimony of Pianki both plead in favour of a walled city.

Each year the inundation began just after the summer solstice. Through the single gate in the walls, a busy population came and went after having filled with various foodstuffs all the storage chambers in the city. Then the necessary arrangements were made to block up the entrance. The wheel had just been invented, as we can see from a bas-relief (Figure 86) showing a siege tower mounted on wheels dating from the fifth dynasty, the period when Memphis was constructed and the building of pyramids was slowing down. It is not extravagant to claim that, to gain time after a number of disasters, a special gate was invented, built in advance and mounted on wheels, that was rolled into position and sealed at the last moment before the water started to rise.

Time continued its steady tread, driving the clouds over the highlands of Ethiopia and provoking the monsoon, whose heavy rains poured down the Nile each year towards Memphis, the giant city charged with history and drama. How many times did the ramparts give way? And when was it finally decided to give up and allow the Temple of Ptah to be annexed by the water? No one can say exactly. Nevertheless it can be argued that such disasters became rarer because the Nile gradually became less impetuous and man's knowledge of hydrology became more extensive. Moreover, the Egyptians of the time, as the heirs of the builders of the great pyramids, were not discouraged by stonework. It is easy to show that, even taking account of the final dimensions of the city, the volume of masonry required for the wall did not exceed half that required for the Great Pyramid; it was not a superhuman task. And yet it called for enormous skill to combat the extreme slyness of infiltrating water.

The city was founded around 2 400 BC. Towards 450 BC, Herodotus paid many visits to the temple priests of Ptah, reputed for their learning, to gather information for his *Survey*. Later, in the second decade before our era, Strabo visited the city and found that it was still lively; at the time it had existed for almost two and a half millennia.

Figure 86 ▶

Siege tower on wheels dating from the fifth dynasty.

Was its demise a consequence of religious quarrels? Was it the Roman Emperor Theodosius I, who, around 380 AD, forbade all pagan rites and thus ended the life of the city after nearly three thousand years of existence? To ease the task of those ordered to destroy the city, did he issue the command not to close its gate at the time of the inundation, allowing the river to invade and destroy Memphis? Or was it, more simply, a victim of the Nile's extreme anger? There was indeed one gigantic flood, in 742 AD, which destroyed the cities along the western branch of the delta, in particular those now under the water in Aboukir Bay which had previously lain on the banks of the Canopic branch of the delta (see Figure 92). The glorious city of Memphis has to some extent become a legend, like Ys.

The drowning of Lower Nubia in 1964 by the damming of the Nile

In most cases the drowning of a civilization is caused by tectonic movements or by extreme weather conditions against which man is powerless, or by human clumsiness as in the case of Venice. But the drowning of Lower Nubia was different, being the consequence of Nasser's dreams of grandeur shortly after the liberation of Egypt and of a design error in the development plans for the river that had nourished Egypt's history.

The young dictator had taken power and, to accomplish his giant projects, had turned to the Soviet Union. It was a fundamental mistake: rivers are like human beings. *Essere fiume*, the nature of rivers, as Giuseppe Penone wrote. You must learn to know them well. The Nile descends in steps towards the plain, which it inundates with its fertilizing waters; to augment its benefits, the best thing to do was to add a few more steps of the same kind.

But Nasser had wanted to impress public opinion. He had decided that the great plain of Egypt, deprived of its annual inundation, would henceforth lie at the foot of a gigantic pile of stone and rubble blocking the river at Aswan and holding back some 168 billion cubic metres of water in a lake that would stretch southwards for some 500 kilometres, of which 200 km were in Sudanese territory.

◀ *Figure 87*

The High Dam (with the upstream lake to the right) and the head of its creator: a document of Nasserian propaganda.

The charming valley of Lower Nubia was condemned to vanish beneath the water. The older Nubians were perfectly aware that Al Qahira, the immense city to the north, had only existed for a thousand years or so, whereas certain of their homes were five times as old. They could not understand why so many foreigners rushed to save temples that sheltered no one rather than the infinitely more precious homes of their families.

The Nubians resigned themselves to their fate and, as they left their valley, asked their ancestors, whose tombs would be submerged by the lake, to pardon them.

Figure 88 ▶

Nubians bringing their gold to Pharaonic Egypt. No doubt Nasser considered that the time for gratitude was over.

The diagram below (Figure 89) shows the comparative levels of the sea, the base of the Great Pyramid and the surface of the lake representing Nasser's dream. The considerable number of dams already built down river from Aswan should have invited him to give serious consideration to a harmonious descent of the Nile in gradual steps. In the event of war, how dangerous it was to place all one's reserves of water in the same lake! Custom required that, in the event of a catastrophic flood, like certain inundations at the time of the pharaohs (which were seen as fits of divine anger on the part of the river), an overflow should channel the excess water towards the upper part of the valley after the impetuousness of the current had been tamed by stones shaped like teeth.

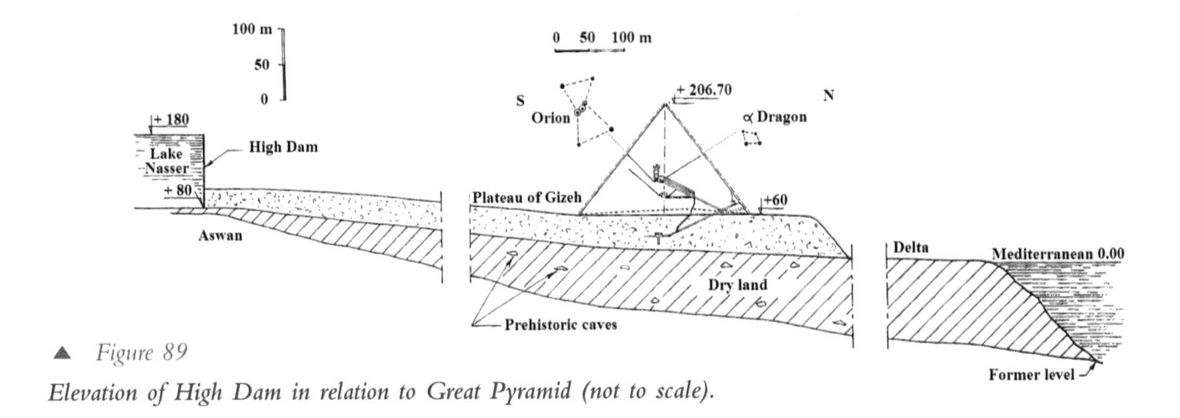

▲ *Figure 89*

Elevation of High Dam in relation to Great Pyramid (not to scale).

Unfortunately, the spillway for excess floodwater designed by the Soviets proved to be faulty. Nasser died long before the reservoir was full; indeed, the time it took to fill was a measure of its gigantic proportions. Nasser's successor, Sadate, considered that in future, in peacetime but above all during a war, 168 billion cubic metres of water poised over a plain concentrating all the wealth of the country was a real danger: the entire civilization could disappear in a few instants. He ordered a study to find a more rational solution for the evacuation of very big floods. The lack of culture of the authors of the project became evident: the Soviets had as little idea of the history of Ancient Egypt as they had of Ancient Mesopotamia. In Egypt, the pharaohs had steered excess floodwater towards Lake Moeris, a depression to the west of the Nile valley; if only the Soviets had taken the pains to read the words of Arrian recounting the history of the Pallacopas in Mesopotamia, they would have learned that "*The Pallacopas lies at some 800 stadia from Babylon.... As the current of the Euphrates carries a considerable volume of water in spring and summer, it bursts its banks and spreads over the land of the Assyrians.... It would flood the country if the people did not make a breach in its banks to let the water flow into the Pallacopas and be diverted towards the marshes and lakes that begin at this canal*" (Arrian, Anabasis, VII, 21, 1-6).

A few old treatises on geology and geography were consulted and it was discovered that the massif of Nubian sandstone, along the western rim of which flowed the ancient Nile, had shifted the bed of the river to its present position above a vast dry depression. The depression of Toshka, some 250 km south of Aswan, was large enough to store the water of catastrophic floods even if they occurred in several successive years. There is a stela in the area marking a spot where extremely hard stone was quarried for the pharaoh Chephren.

The dam should have been built opposite this depression at Toshka, 250 km further upstream. It needed to be only half the height of the present dam and to be equipped with a lateral spillway for major floods, which would have protected the Nile valley from every possible caprice of the river. How difficult it is, even today, to admit that Nasser made a mistake!

I am reminded of the opening words of Herodotus' History, written in about 450 BC: "*Herodotus of Halicarnassus here displays his inquiry, so that human achievements may not become forgotten in time, and great and marvellous deeds – some displayed by Greeks, some by barbarians – may not be without their glory*". The barbarians of the south, as the Egyptians were called by the Greeks, had gathered in the valley of Lower Nubia a rich testimony of these "*marvellous deeds*" and erected more than twenty temples. The stones of those temples told of the doings of all the glorious pharaohs who had defended their country against invasions from the south or established religious foundations. There they were, in their own universe, like "*fruits of the invisible tree of time*". Some had a special relationship with the river or, as in the case of the temple of Abou Simbel, with the rising sun. One could almost hear the commanding voice of Sesostris III, builder of fortresses, or of Ramesses II proclaiming himself lord of the universe and wanting to efface for ever the memory of Akhenaton: long before Herodotus, the stones had commemorated many lives passed in the service of illustrious pharaohs.

Rapidly and generously about a dozen nations came to the rescue of temples and monuments doomed to be submerged on the orders of Nasser. Many temples were saved but the flooding of the valley of Lower Nubia obliterated a large chunk of human history. Traces of a human presence disappeared into the gigantic lake; caves once inhabited are now deep under water or filled with the silt that is rapidly invading the bed of the reservoir. Admittedly, there still remain above the water line a few buildings that remind us of the drowned valley's past: they are the stones of temples that have been dismantled and then reassembled, tokens of the international rescue operation. Some of them were given to foreign nations as a sign of Egypt's gratitude but, scattered as they are in all corners of the world in sterile air-conditioned museums, they have lost the ability to speak of the past and the glory of the pharaohs.

The absurd project for a second valley

The wrong site for a dam; unfortunately this error is about to be compounded by another. Instead of regarding Toshka as an *occasional* means of dealing with excess floodwater, the Egyptians have come up with the idea of using the depression as the starting point for a second valley that will join up the string of distant oases. This project is quite rightly condemned by all the countries further upstream: Egypt already consumes more water than was attributed to it in the most recent treaty with Sudan. The water that pours into this second valley will run to waste in the burning heat of the desert. This thermodynamics of the absurd will be punished one day by the upriver countries: in response, they will tap the river at its source.

The land possessing the headwaters is always in the most powerful position: if it is not so in this particular case, it is because of the British occupation long ago and the backwardness of Sudan and Ethiopia. These two countries are now showing signs of rebelling. The stela of Aksum, *emblem of black Africa*, embodies a spirit like that in the distant past when the Abyssinians were, for all the peoples of the Nile, "*the masters of the waters*", those who could block its course or make it flow in the opposite direction. For the Muslim Pasha of Cairo, the very Christian Negus was an enemy who had to be treated with respect on account of his power over the Nile. A fourteenth century Negus of Ethiopia wrote: "*The waters of the river that rises in my country will be stopped from reaching yours, which I shall cause to die of thirst*".

Whereas Egypt is in the midst of an unprecedented economic crisis, Sudan's new oil resources are enabling it to start work on two dams north of Khartum for irrigated agriculture, in particular for the growing of cotton, which consumes great quantities of water. They will deprive the Egyptians of part of the Nile's discharge. Make no mistake, these two Sudanese dams are merely the advance guard of a campaign to regain control over the upper reaches of the Nile, like the twenty-two projects implemented by the Turks on the Tigris and Euphrates, which have parched Syria and Iraq. Not only will they show the stupidity of the second valley but they will also cause a problem of deep concern in Egypt: the fine stones of the Great Pyramid will no longer be reflected in the waters of the Nile. Eventually the historic river will flow through a drier valley in a narrower bed, carrying its dirty water through a desert of sand.

The Suez isthmus: meeting point of three tectonic plates

The isthmus has a long history. In the first two millennia before our era it facilitated relations between the biblical peoples and Egypt and served as the route for invasions by Babylonians, Assyrians and Persians. It watched the tramp of the armies of Ramesses II, Cleopatra, Alexander the Great, various Roman generals, Saladin, Napoleon Bonaparte and, more recently, the troops of the Six-Day War and the war of October 1973.

Here the depths of the earth are shaping the surface. This historic isthmus is exceptional from the tectonic point of view, being the point at which gigantic opposing forces – the African plate and the Eurasian plate – are in collision and are driving against Europe at the rate of 7.1 mm per year. From time to time, an undersea volcano even reappears in the form of Ferdinandea Island, just south of Sicily. At the same time, further east, the African plate is scraping the Arab block from south to north (Figure 91), creating a shear fault along the edge of the Sinai which is vigorously resisting this shearing and forcing the isthmus to pivot.

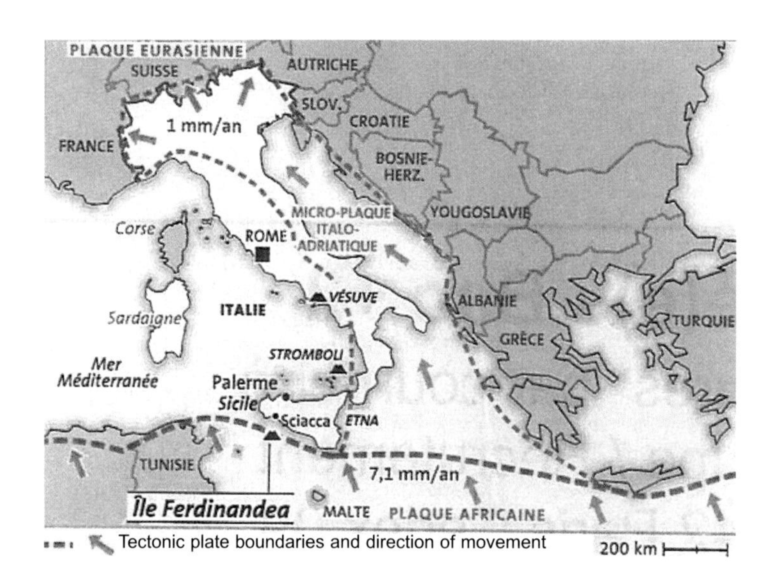

◀ *Figure 90*

Beneath the delta gigantic forces are meeting – the African plate pushing against the Eurasian plate and the shearing of the Arab block.

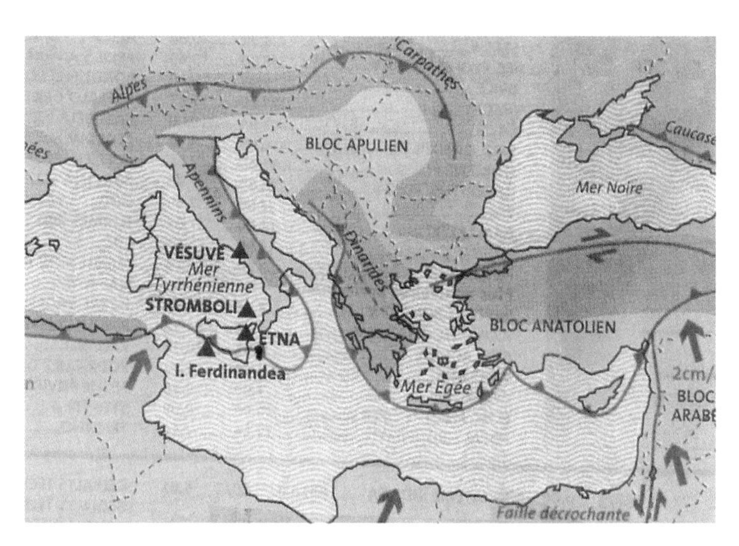

◀ *Figure 91*

Note the north-south shear fault to the east of the delta.

Surrounded by all these colliding forces, the stone has become plastic, rising in some places and sinking in others, the process being accompanied by violent earthquakes. It is therefore hardly surprising that the explosion of the island of Santorin, felt as far away as Alexandria, and, two thousand years later, the earthquake of 21 July 365 AD, which generated a terrible tidal wave that drowned over five thousand people in Alexandria, should have for so long haunted the memories of that city's inhabitants.

Clearly, the eastern branch of the Nile delta, the so-called Pelusian branch, has dried up: Pelusium[60], formerly a city of mud, has become a tell, a dry mound; Per-Ramesses has been abandoned in favour of Tanis to the west. Conversely, Rosetta, founded in the ninth century, which used to be the principal port of Egypt, is now no more than a halt for coastal traffic. The level of the Canopic branch of the Nile, which used to be the most westward one, has subsided by several metres with the result that the mouth of the river itself has disappeared and created Aboukir Bay. There appears to have been a tilting movement with the area around Sais serving as a pivot (Figure 92). In the port of Alexandria, west of Canopus, rocks that used to break the surface in the time of Strabo have disappeared under water, confirming our hypothesis that the delta has tilted sideways.

In short, the level of the land has varied considerably over time, with the eastern part of the delta being uplifted and the western part subsiding.

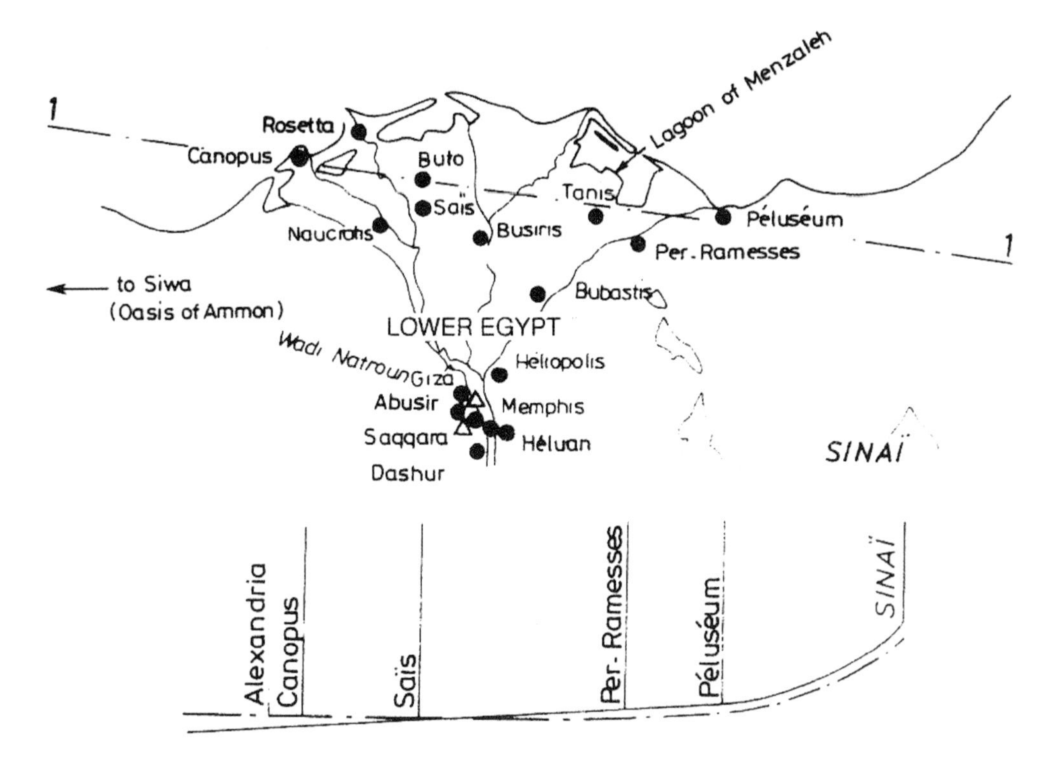

▲ *Figure 92*

The Nile delta, showing how it has pivoted. Former level: —— - —— - Present level: ——

[60] *"Pelusium is surrounded by marshes that some people call Barahra and by stagnant pools of water"*, Strabo, 21.

The delta as it is today is the result of two phenomena – a tilting movement and, over the last fifty years or so, a retreat. Guiseppe Penone states in his book *Essere fiume* (The nature of rivers): "*The river transports the mountain. The river is the carrier of the mountain*". This is no longer true of the Nile: the High Dam is like an immense wine skin that hoards the particles torn from the highlands of Ethiopia.

Fragments of an unknown Egypt

In the murky water of the port of Alexandria, several teams are engaged in underwater archaeology in a number of places. The Ptolemies were not the most orderly of rulers, as I have stressed, and the remains of different civilizations are often mingled, which can be frustrating for archaeologists working under water.

Figure 93 ▲
A gold ring found between two ribs of a wrecked ship in the royal harbour of Antirhodos.

▲ *Figure 94*

The head of Caesarion stripped of salt water concretions. Caesarion was the ill-starred son of Cleopatra VII and Julius Caesar.

Figure 95 ▲
The island of Pharos and the Ottoman fortress built by the Sultan Qait-bey.

Most Egyptologists think that the fortress of Qait-bey was built on the site of the ancient lighthouse, at the end of the heptastadion. This is far from certain.

Heracleion and Menouthis

The Canopic mouth of the Nile delta (Figure 92) was one of the most commonly used in former days. It was the one used by Herodotus on his way to Naucratis, in 450 BC the only port open to Greek ships wanting to enter Egypt. Herodotus speaks of two mouths at each extreme of the delta, the 'Pelusian' in the east and the 'Canopic' in the west (II, 15) and observes (II, 5) that "*if you take a cast of the lead a day's sail off-shore, you will get eleven fathoms, muddy bottom – which shows how far out the silt of the river extends*". For him, Egypt was a land of alluvium, "*a gift of the Nile*".

Along that mouth existed two cities, Heracleion and Menouthis, founded in the seventh and sixth centuries BC, during the reign of Nectanebo I. Shortly afterwards, in 331 BC, Alexander founded the city that bears his name on the shores of the bay some 24 km west of those two cities. Strabo, in around 24 BC, described the Canopic mouth and mentioned the cities of Heracleion and

Menouthis: "*Crowds of people would come to Canopus from Alexandria for public holidays: men and women would dance outrageously, in the most lascivious manner, with the population of Canopus, who have houses beside the canal well placed for festivities of that kind*". The Canopic towns were reputed for their relaxed lifestyle: Anthony and Cleopatra went there to celebrate their love discreetly. Heracleion, at the mouth of the river, was a prosperous customs town while Menouthis lived mainly from the pilgrimages to its famous Temple of Isis, perhaps a desire to make amends for its sinful past.

But the fact remains: at some unknown period[61] after the time of Strabo, the extremity of the Canopic branch collapsed and created a bay shaped like a large bite out of the coastline. It is called Aboukir Bay. One consequence of this collapse was the disappearance of the two cities. It was then but a small step to maintain that the Creator had wanted to drown the cities as if they were another Sodom and to pour scorn on the idea that the Canopic mouth of the Nile had been increasingly destabilized and eventually engulfed by a tilting movement of the land. Indeed, Strabo himself had something of the geologist when he wrote (Book 36): "*It must be admitted that a large proportion of the land turns into water, and vice versa...*"!

Franck Goddio has extended his underwater explorations to this part of Aboukir Bay with brilliant success: in 1998 he discovered, at a depth of six metres, two kilometres from the coast, the remains of Menouthis and, six kilometres from the coast, those of Heracleion, the latter covering an area of one square kilometre. There have been some fascinating finds which, taken together, are of considerable importance. They would fill several rooms of a museum and make many other Egyptologists, who might once in a while discover a tomb, pale with envy. As the Secretary General of the Supreme Council of Egyptian Antiquities enthusiastically remarked: "*We are talking about whole towns that are mentioned only in ancient texts*".

Franck Goddio is a controversial figure: he is criticized for using special funds and exploiting his work for commercial ends both here and elsewhere, including the Philippines and the China Sea, where he has found some galleons and brought amphora and plates to the surface at a time when UNESCO is endeavouring to protect wrecks and treasures lying at the bottom of the sea. But how could he maintain a ship and its crew, and all the necessary electronic equipment, without money? The jealousy of certain persons even leads them to give credence to certain press reports suggesting that the past of Goddio is not as limpid as the waters in which he carries out his research. Charming colleagues! Alone, Jean Yoyotte, whose competence is unquestioned, is untouched by this trend. Moreover, there is no doubt that the objects found in the excavations have been placed at the disposal of the community.

After examining the site, Yoyotte concluded that the two cities had expanded to the point where they had merged into a single one covering an area of 120 hectares. One can see a temple and docks lined with walls of large limestone blocks; numerous objects dating from the first centuries BC have been excavated, showing that the city was prosperous at the time of the early Ptolemies. As Yoyotte has pointed out, one of the most extraordinary finds was a stela inscribed with the decisions taken by Nectanebo I in 380 BC in favour of the temple of Neith at Sais, an exact copy

[61] After the Umayyad Califate in the seventh century, since coins from that period have been found.

of the so-called Naucratis stela which was discovered at Naucratis in 1890. Both stipulate that 10% of the customs dues paid by Greek merchant ships should be given to the temple of Neith.

Landslide or earthquake? Various analyses of the delta's sediments reveal that the Canopic branch shifted eastwards before moving back towards the west and destroying the two cities. The date of the most recent coin found on the seabed, an Arabic coin of 742 AD, coincides with the year of a major inundation.

◀ *Figure 96*
A magnificent head in white marble of Serapis, ptolemaic period. Copyright © Franck Goddio/Hilti Foundation. Photographer Christoph Gerigk.

▲ Figure 97

Royal head of a statue in diorite, XXVI Dynasty (654-525 BC) found during the archaeological excavation in Aboukir Bay on the site of the submerged suburbs of the city of Canopus.

Copyright © Franck Goddio/Hilti Foundation. Photographer Christoph Gerigk.

Figure 98 ▲

Gold coins of the early Islamic and Byzantine periods unearthed from the western site, Aboukir.
Copyright © Franck Goddio/Hilti Foundation. Photographer Christoph Gerigk.

12

Stone for the eternal home of the Pharaohs

The world under our feet has always fascinated man. First as a refuge: "*and, lo, there was a great earthquake ... and the kings of the earth, and the great men, and the rich men ... hid themselves in the dens and in the rocks of the mountains*", it is written on the Book of Revelations (VI, verses 12-15). On the other hand, the American Indian prophet Smohalla[62], of the Umatilla tribe, utterly refused to make the least incision in the earth: "*It is a sin*", he said, "*to wound or cut into, to tear or scratch our common mother with our hands. Shall I take a knife and plunge it into my mother's bosom? Shall I dig under her skin to reach her bones? Then when I die, I cannot enter her body to be born again*". If we exhaust the earth it will take revenge. When the very ancient Egyptians themselves exploited quarries or mines, they felt that they were purloining the possessions of the legitimate owner of the site. The mountain was sacred: it was the land of the god Min or of a goddess and, to obtain their forgiveness, they would engrave a stela or burn incense.

Later, however, an astonishing variety of very different solutions was adopted in many countries[63]. In the fifth millennium BC, for example, near the Atlantic coast, but also in China, a common practice was to build tumuli, great mounds of stone and sand covering and concealing a burial chamber in which the body of an illustrious personage lay under a shroud of earth. The largest of these mounds (150 m in diameter and 40 m high) is in England, at Silbury, east of Bristol.

In China, certain emperors cherished the belief that death would spare them, especially if they hid themselves under the earth. The founder of the first imperial dynasty, Qin Shi Huangdi (259-210 BC), a bloodthirsty tyrant, aspired throughout his life to the one thing he could not have – immortality. To achieve his goal, he ordered the construction of a tomb that resembled a fortress. Fearing that all those he had unscrupulously eliminated might trouble his rest, he had chosen for his burial the centre of a large pyramid which was partly below ground level. Around it were buried eight thousand warriors in pottery, armed to the teeth. He himself believed he could vanquish invincible death by

---

[62] See James Mooney, *The Ghost-dance religion and the Sioux outbreak of 1890* (Annual Report of the Bureau of American Ethnology, XIV.2, Washington, 1893, pp. 641-1136).

[63] See J. Kerisel, *La pyramide à travers les âges*, Paris, Presses de l'Ecole Nationale des Ponts et Chaussées, 1991.

▲ *Figure 99*

Civilization of the Nabateans.

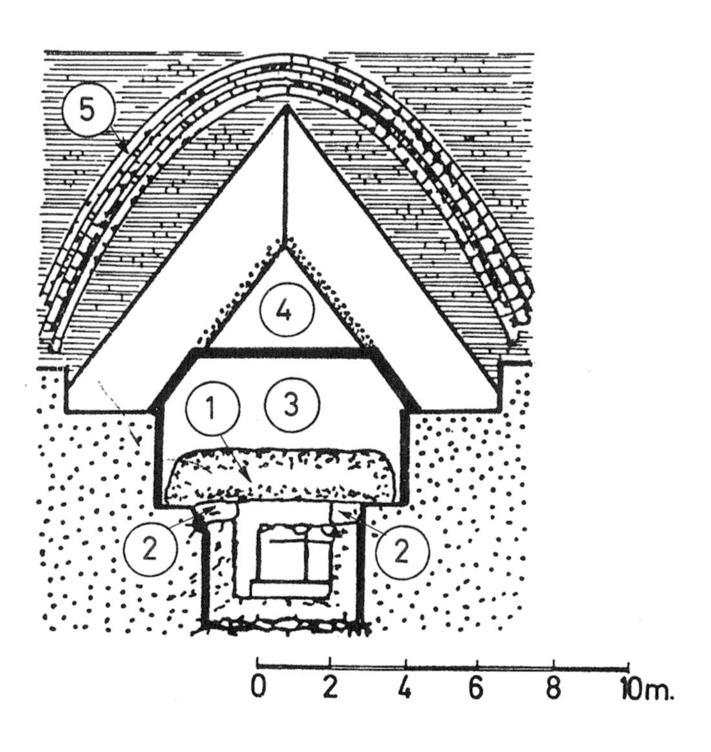

Figure 100 ▶

Burial chamber of Amenemhat III at Hawara.

1. Three quartzite beams
2. Side walls
3. Protective chamber
4. Stone rafters
5. Relieving vault

swallowing magic herbs. It is only recently that this strange world of an underground army of a power-hungry emperor has been discovered.

The Egyptians invented the mastaba, half buried in the ground and containing earthly foods to reassure a person who believed in a second life and to sweeten its course. Then the pharaoh Djoser invented the pyramid. But the size of the pyramid was not reflected in the size of the burial chamber: sometimes the enormous pyramid appears to crush, as it were, the tiny tomb. The Great Pyramid and its burial chamber were exceptional cases, clearly the work of a genius.

One of the last burial chambers placed under a pyramid was that of Amenemhat III at Hawara. With its excessive emphasis on security, the design is distressing: the pharaoh is walled into a tiny space with several different systems of protection – but no decoration of the walls.

At the time of the pharaohs, Egypt had many more earthquakes than nowadays and its rulers, mistakenly fearing that their pyramids might collapse, decided to spend their second lives resting in the bowels of the earth in the Valley of the Kings.

The tomb of Ramesses II

The man in the street in Thebes did not have to be a learned geologist to realize that the low-lying Valley of the Kings was not without danger since it was underlain by the famous "*Theban shale*", which is very sensitive to outside agents and especially water.

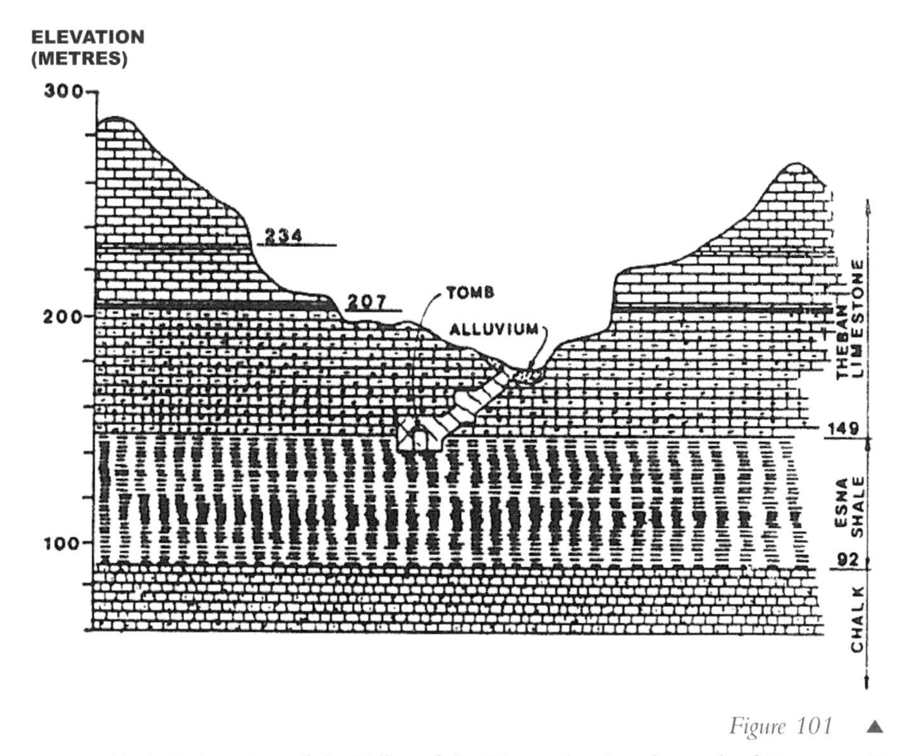

Figure 101 ▲

Geological section of the Valley of the Kings, showing the tomb of Ramesses II.

Ramesses II was too full of himself to pay attention to such things: nature should simply bow to his wishes. The adjective 'pharaonic', so often applied to the works of this pharaoh, conjures up the enormous statues and immense palaces that embodied his desire for lasting fame. Hecateus of Abdera, during the reign of Ptolemy Soter, came to Egypt and had translated into Greek the inscription on one of the statues of Ramesses II at Thebes. It begins: "*I am Ramesses, king of kings... if someone wishes to know how great I am and where I am to be found, let him surpass one of my works*". He exhausted the resources of Egypt's quarries and mines. Fortunate Ramesses: he spent the whole of a very long life squeezing glory from a single battle, the battle of Kadesh, which he had won, as he said, "*with a total victory*" in the fourth year of his reign. The battle in question was most probably a simple skirmish with the Hittites on the banks of the Orontes. No text from a different source, however, has ever been found to clarify the matter. After being regarded, thanks to the Bible, as the quintessence of evil, he has now been largely rehabilitated owing to this feat of arms.

Ramesses II had chosen for his tomb in the Valley of the Kings the lowest site in the valley, on the right as one enters it. In death he was determined to be the "*first among equals*" and his tomb would be the very first at the entrance to the valley. The site he chose contrasted with the much more discreet location at the head of the valley chosen by his predecessor Thutmosis I, a valiant warrior and the first pharaoh to select that valley for his eternal dwelling-place. The pride of Ramesses and his desire to be pre-eminent would prove fatal.

The forces of evil dwelt in the chosen site: or, more precisely, the geology and hydraulogy of that particular spot were most unfavourable[64]. Whereas all the other tombs in the Valley are cut into

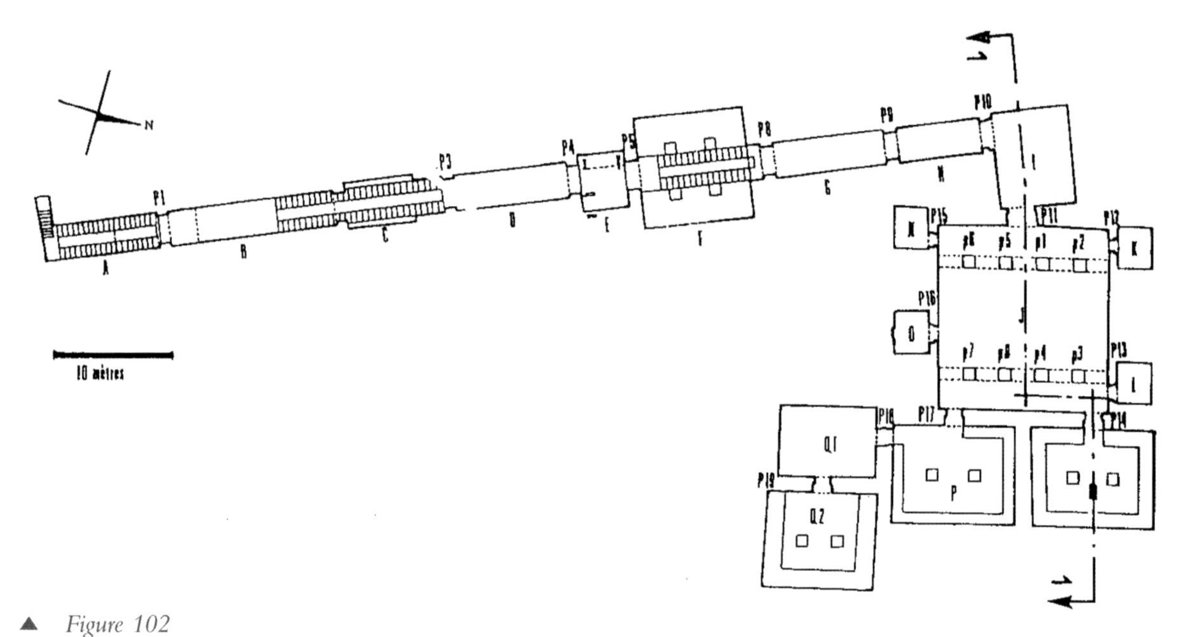

▲ *Figure 102*

Plan of the tomb of Ramesses II, based on measurements by the Theban Mapping Project.

[64] A detailed study of the processes responsible for the gradual destruction of the tomb was presented by G. Curtis and J. Rutherford to the Tenth International Congress of Soil Mechanics in 1981.

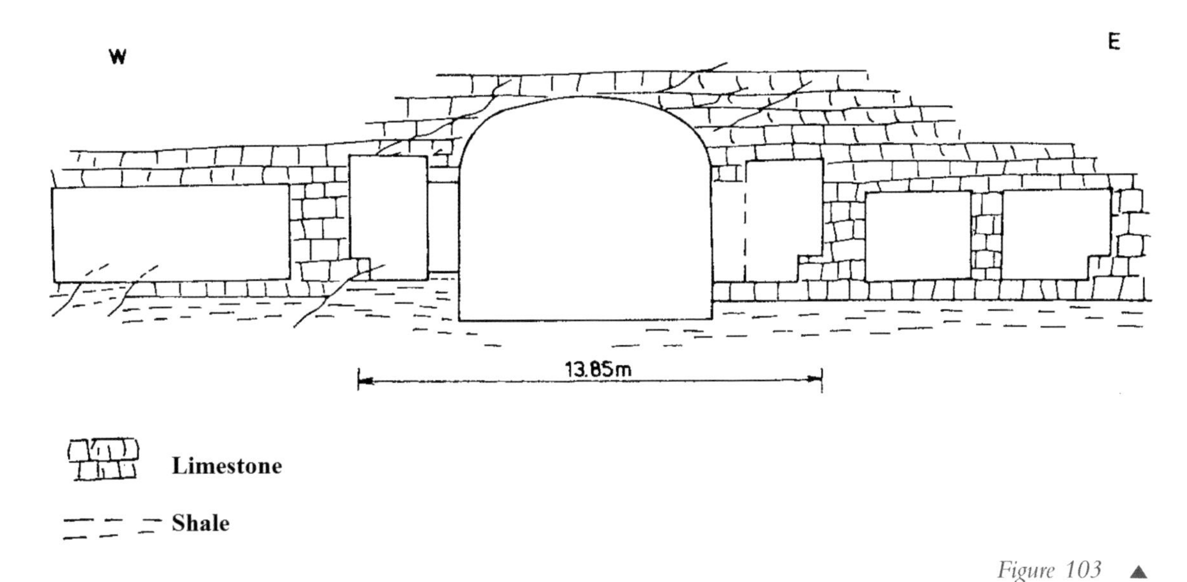

Figure 103 ▲
Vertical section along the west-east axis 1-1 in Figure 102.

the mass of the Theban limestone, a first-rate stone, in this case, because of the low elevation of the site, the floor of the large chamber lies at the level of the underlying Esna shale. When this shale is imbibed with water, it swells and exerts a very considerable pressure: this is precisely what occurred, for rainwater seeped down through the ceiling of porous limestone. Even the lower part of the Theban limestone, which forms the roof of the chamber, was contaminated by the underlying shale, which made it swell under the influence of the humidity oozing in after storms. In other words, the roof of the chamber was like a leaking umbrella.

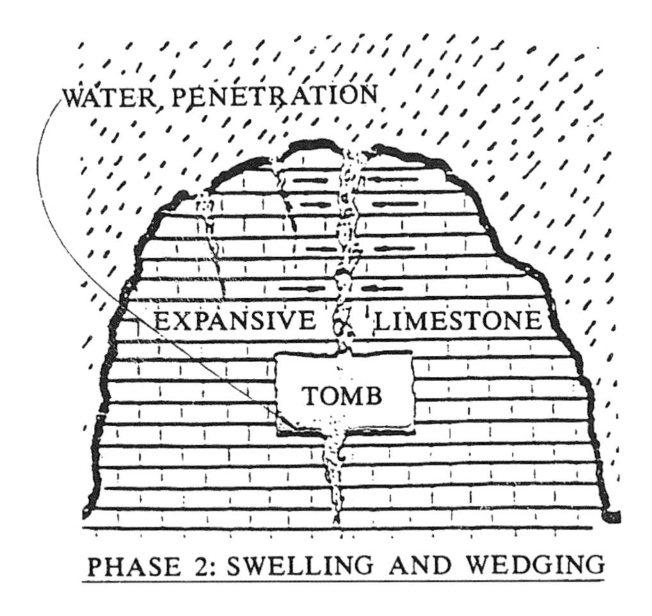

◀ *Figure 104*
The tomb of Ramesses II, under a leaking umbrella.

Thus the tomb of Ramesses II has become over the last thirty years a classic example of errors to avoid in applied geology (Figure 104). The shale on which the pillars were founded reacted to the humidity by exerting enormous pressure, forcing the pillars upwards into the vault. Conversely, during the dry season, the underlying shale retracted and, in descending, pulled at the pillars: a *danse macabre* within the tomb of the great Ramesses II.

The burial chamber was of a noble design: a large room with a basket-handle arch supported by eight pillars, with side chapels that foreshadow the ambulatories of our churches. But such slender supports could not withstand the pressure of the vault. Whereas the burial chamber in most tombs has only six pillars, this one has eight, as if Ramesses had a premonition that he was risking a dangerous challenge to the laws of soil mechanics.

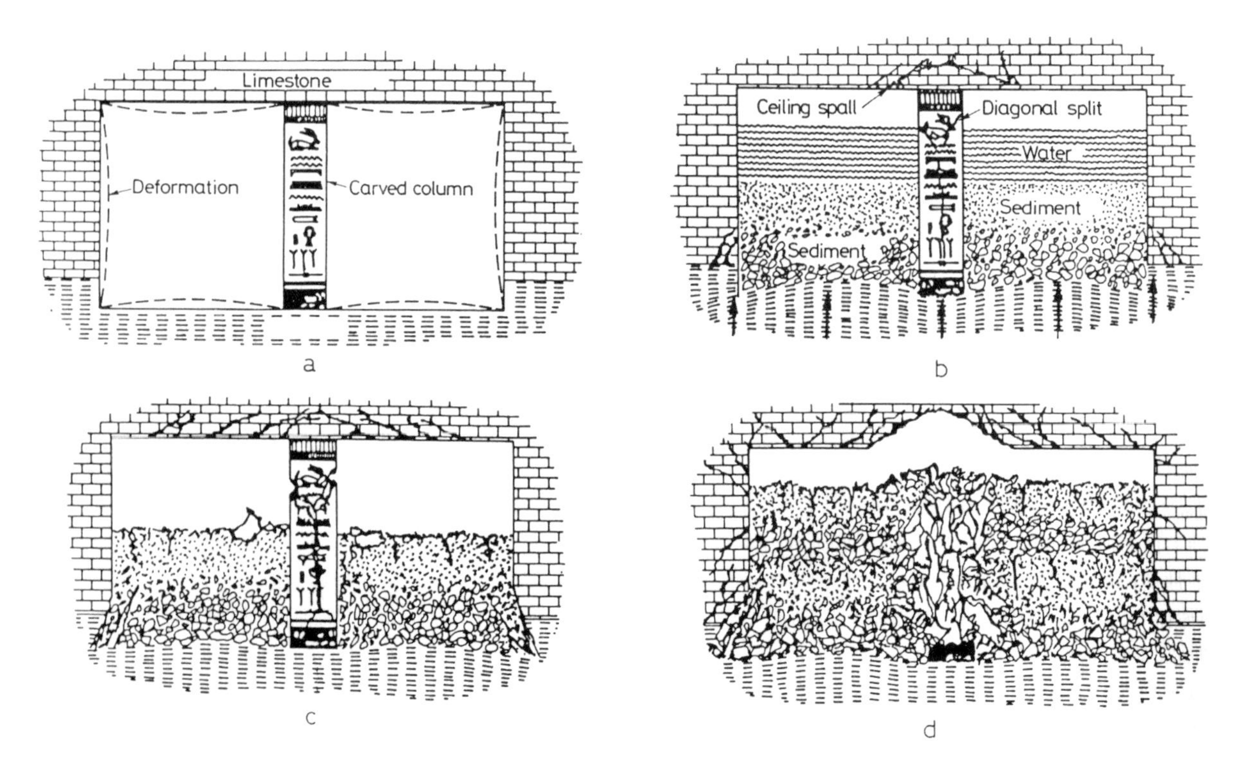

▲ *Figure 105*

Vanitas vanitatum! The successive tortures inflicted on the burial chamber of Ramesses II by natural forces (Curtis and Rutherford, Brooklyn Museum, 1981).

Phase a: Initial excavation. The weight of over 280 m of rock on the roof of the chamber has resulted in some minor distortion and the concentration of forces at the corners.

Phase b: Flooding and swelling. The saturated shale swells and exerts enormous pressure on the pillars and partitions, splitting walls and columns.

Phase c: Drying out. The drying shale slowly contracts, leaving portions of the pillars broken and walls hanging from the roof structures.

Phase d: Collapse. After several cycles of flooding and drying, the ceiling crumbles and falls.

After having been looted – the supreme insult – the tomb was almost entirely filled with mud brought in by occasional torrential rains: detritus and gravel invaded the tomb, which became so to speak the cesspit of the Valley.

Curious fate for a brilliant pharaoh in love with fame. Archaeologists will find this account unseemly for such a great pharaoh, but it is a sad truth that archaeologists and geologists all too often ignore each other.

As a person passionately interested in Cheops, the other great pharaoh, I was naturally drawn to compare their respective conceptions of the second life. For Cheops life and death are part of the same current, but at opposite extremes : on the left is life, represented by the buoyant water of the canal serving the pyramids, and on the right the water of the funerary canal on which, when the time came, a bark would bear the pharaoh towards a new life of light and place his mummy on a platform of rock surrounded by the water of the Nile, as Herodotus reports. Life, death, the water of the historic river – there is much grandeur in this thriad.

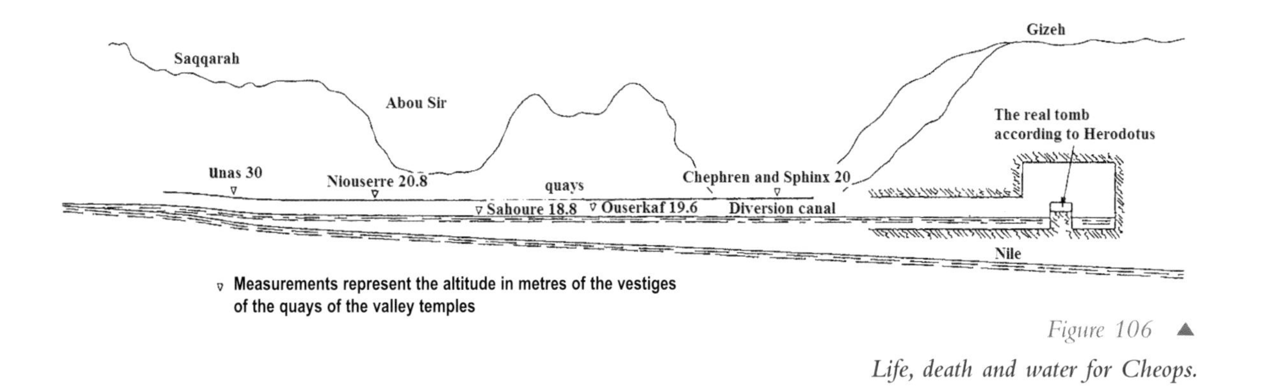

Figure 106 ▲
Life, death and water for Cheops.

It was said that Ramesses II would be spared nothing: after his death, he had the impression, as the water seeped in, that invisible hands were lifting him up and then laying him down again. In the twenty-fifth year of the reign of Ramesses IX, his mummy, in danger of being looted, was taken out of its sarcophagus and subsequently transported by the High Priest Pinedjem, in about 970 BC, to the celebrated hiding place at Deir el-Bahri, which remained inviolate until the nineteenth century. A third voyage then took him to the Museum of Cairo, a fourth to Paris and a fifth and last brought him back to Cairo. Poor wandering soul!

Why these tribulations? At Deir el-Bahri numerous other mummies were kept in a humid atmosphere and, over nearly three thousand years, the hiding-place was a feast for mushrooms. Brought back to Cairo, Ramesses II learned to his cost that a life, and even a second life, is never over: thousands of men and women, if only to bask a little in the reflected glory of his reign, take an interest in him that is impossible to prevent. Thus Ramesses came to Paris to have his fungi treated – without much effect as the treatment was not carried out in a sterile environment. The mummy was contaminated and remained so; the researchers had to wear a mask, but not the mummy. We learned, if we did not know the fact already, that he was nearly six feet tall and had red hair.

Today he reposes in silence in Room 56 of the Cairo Museum, a little thinner after his X-ray treatment in Paris, alongside a few "*colleagues*" whose identification had been relatively easy. The fact is that, for many other mummies, doubts remain as to their identity because of accidental swapping of the "*identity papyrus*" in the event of transfer or theft. The ADN is unexploitable but certain facial or cranial measurements under X-rays can be used to identify certain individuals or blood relationships.

It appears that not all these mummies have been subjected to serious mycological research. On the other hand, the scientific analysis of some Coptic mummies, by Dr Ezzedine Taha and Dr Abd el Elaal, has demonstrated the presence of *Aspergillus niger*, among others, thus proving the existence of human or aerial contamination.

Welcome back to your country, Oh great Ramesses, and may you now be left in peace!

A small team of researchers from the French National Centre for Scientific Research (C. Leblanc and A. Guillaume), in association with the Louvre Museum and initially with the backing of the Elf Foundation, has undertaken to clear and restore the tomb of Ramesses II. It is intended to install a museum in it to satisfy the curiosity of the living: under the restored vault it will be possible this time to breathe in the scent of immortality and reflect on the ambitions and errors of the great King.

The pharaoh had fifty-two sons and the tomb built for them on the other side of the entrance to the valley is of considerable size, with 110 small chambers off an immense hall of 325 m^2 supported on sixteen pillars. Because this immense hypogeum is located at a higher level, it has had a happier fate.

Final thoughts

Roger Caillois looked at stones with feelings of admiration. For him they were emblems of longevity: stones existed long before the first human being came on the scene and they would still be there when all else was destroyed. "*How can we not venerate them? In comparison, everything else – man and his works - is fragile and transitory. For whom should I speak? There is no need to choose*"[65].

In this book, we have simply wanted to show the ungratefulness often displayed by mankind when realizing his works of stone, to listen to the grievances of stones as they groan in the wind at the pinnacle of man's architectural works and to denounce their pointless submersion. As Octavio Paz observed in speaking of Roger Caillois: "*We do not read in the same way as we look at a stone. As we examine a text, there comes a moment when our act of reading is transformed into something else, something that does not negate it but rather complements it – contemplation. At this moment we are reading just like Roger Caillois used to read the signs engraved on each stone – as the echoes and reflections of an incorporeal time*"[66].

We must also return to Herodotus, who wrote in the preface to his magnificent History: "*Herodotus of Halicarnassus here displays his inquiry, so that human achievements may not become forgotten in time*". The "*father of history*" had already sensed this erosion, sometimes active sometimes less obvious, in the course of his travels. By the time of his return to Thurii, he knew that the comprehensive account of the known world that he was going to write was already out of date. He sensed the passing of time: "*There was a time when the Egyptians had no country at all... they have existed ever since men appeared on the earth, and as the Delta increased with the passage of time, many of them moved down into the new territory and many remained where they originally were*" (Book 2, 15). Remarkable vision of a past in movement. This past in movement, which can be perceived in stones, silt and sand, remained to be explored – and that is what we have tried to do.

In 1915, Freud, at a time when great segments of our European culture were collapsing, wrote: "*A time may come indeed when the pictures and statues which we admire today will crumble to dust, or a race of man may follow us who no longer understand the works of our poets and thinkers, or a geological epoch may even arrive when all animal life on earth ceases; but since the value of all this beauty and perfection is determined only by its significance for our emotional lives, it has no need to survive us and is therefore independent of absolute duration*"[67].

We do not think that that day will come, and the word "*need*" is shocking. Of course, we are

---

[65] Quoted by Octavio Paz in his address to UNESCO on 13 May 1991 on the occasion of an international meeting to honour Roger Caillois.
[66] Ibid.
[67] Sigmund Freud, On Transcience, 1916.

irremediably dispossessed by the work of time or by the barbarity of people fighting each other. Freud contrasts our "*Wunschleben*" – our desire to live – with "*Realitätwerke*", or harsh reality.

But, Dr. Freud, reality depends in the end on ourselves: civilizations are changing constantly, as Herodotus observed, and so are mental outlooks. *Mens agitat molem* – the mind enlivens the whole mass – is still true, but the mind that enlivens matter is now more intelligent and we now know more about matter. The ancients were interested in the sexology of stones and you, Dr Freud, in that of human beings. Like you, the ancients did not perceive the danger of unnatural marriages in which the male part bore an unacceptable share of the burden.

We now know how to avoid the disasters that could result: the sudden collapse of a bell-tower will not happen again. We will discover how to combat the dangerous inclination of leaning towers and we will discover how to construct even heavier domes without risk. How pessimistic Lucan appears to us with his *ruinae etiam periere - even the ruins will perish*. From a heap of ruins today's archaeologists can extract a fine vision of the past. The progress of underwater archaeology promises to bring to light whole cities that were engulfed by the sea more than a thousand years ago in the same way as brilliant archaeologist-historians have succeeded in finding, under their shrouds of sand, the remains of great civilizations. Rarer and rarer will be the hermits retreating into the desert to live the life of an anchorite and revere the ashes of their ancestors; more and more numerous will be those who will come to a place of memory to take their sustenance "*on the tall grass of fruitful works*".

Bibliography

Airvaux, J., 2001: *L'Art préhistorique du Poitou-Charente*, La Maison des Roches, Paris.
Allègre, C., 2003: *Un peu de science pour tout le monde*, Fayard, Paris.
Caillois, R., 1965: *Au cœur du fantastique*, Gallimard, Paris.
Celant, Germano, 1989: *Guisseppi Penone*, Electa, Milan.
Bernard, A. and E., 1960: *Les inscriptions des colosses de Memnon*, Cairo.
Boreux, C., 1924-25: *Etudes de nautique égyptienne, l'art de la navigation en Egypte jusqu'à la fin de l'ancien Empire*, Memoir published by the Institut Français d'Archéologie Orientale (MIFAO) n° 50, Cairo.
Breasted, J. H., 1906: *Ancient Records of Egypt*, University of Chicago Press, Chicago.
Breuil, H. (Abbé Henri), 1952: *Quatre cents siècles d'art pariétal*, Centre d'études et de documents préhistoriques, Montignac.
Bunau-Varilla, P., 1920: *Panama, The Creation, Destruction and Resurrection*, Mc Bride, Nast & Co, New-York.
Caillois, R., 1939: *L'Homme et le Sacré*, Leroux, Paris.
--,1965: *Au cœur du fantastique*, Gallimard, Paris
Champollion, J.-F., 1824: *Précis du système hiéroglyphique des anciens égyptiens.*
Chateaubriand, François René, Vicomte de, 1827: *Itinéraire de Paris à Jérusalem et de Jérusalem à Paris*, Garnier, Paris.
--, 1947: *Mémoires d'outre-tombe*, La Pléiade (Gallimard), Paris.
Conant, K. J., 1968: *Cluny: les églises et la maison du chef d'ordre.* The Mediaeval Academy of America, n° 77, Cambridge (Massachusetts).
Congrès International d'Etudes du Canal Interocéanique, Proceedings, 1879, Société de Géographie de Paris, Paris.
Coppens, Y., Picq, P., 2001: *Aux origines de l'humanité*, Fayard, Paris.
Curtis, G., and Rutherford, J., 1981: "Expansive Shale Damage, Theban Royal Tombs, Egypt", Proceedings, Congrès International de Mécanique des Sols et Fondations, Stockholm, 10, vol. 3, pp. 71–74.
Déguignet, J. M., 1999: *Mémoires d'un paysan bas-Breton*, An Here, Ar Releg-Kerhuon.
Desroches Noblecourt, C., 1962: *L'art égyptien*, Presses Universitaires de France, Paris.
Desroches Noblecourt, C., 1997: *Ramsès II, la véritable histoire*, Pygmalion, Paris.
Diesbach, G. (Ghislain de), 1998: *Ferdinand de Lesseps*, Perrin, Paris.
Eliade, M., 1976: *Histoire des croyances et des idées religieuses*, Payot, Paris.
Emery, K. O., 1960: Weathering of the Great Pyramid, *Journal of Sedimentary Petrography*, vol. 30, I, pp. 140–144.
Forot, V., 1910: *L'Ingénieur Godin de Lépinay*, Impr. de Roche, Brive.
Freud, S., 1916: On Transience.
Garbrecht, G., 1983: *Sadd-el-Kafara, the World's Oldest Large Dam*, Bulletin 81, Leichtweiss Institute for Water Research, Technical University, Braunschweig.
Gauthey, É.-M., 1771: *Mémoire sur l'application des principes de la mécanique à la construction des voûtes et dômes*, C.-A. Jombert, Paris.
--, 1798: *Dissertation sur les dégradations survenues aux piliers du dôme du Panthéon français et sur les moyens d'y remédier*, H.-L. Perronneau, Paris.
Giddy, L., 1994: *Memphis Survey: état des recherches archéologiques et épigraphiques*, BSFE, 129.
Goddio, F., Bernand, A., 2002: *L'Egypte engloutie*, Arcperiplus, Londres.
Goyon, G., 1971: *Les ports des pyramides et le grand canal de Memphis*, Revue d'Egyptologie, 23, pp. 137–153.
--, 1977: *Le Secret des bâtisseurs des Grandes Pyramides: Khéops*, Pygmalion, Paris.
Guillaume, A., Humbert, P., Sablon, J.-Y., 1997: "Analyse architectonique par modélisation numérique de la tombe de Ramsès II (Vallée des Rois, Louqsor, Egypte)", *Revue d'Archéométrie*, 21, pp. 67–80.
Habachi, L., 1987: *The Obelisks of Egypt, Skyscrapers of the Past*, The American University in Cairo Press, Cairo.
Hancock, G., 1992: *The Sign and the Seal, A Quest for the Lost Ark of the Covenant*, William Heinemann, London.
Hornung, E., *Der Eine und die Vielen.*
Jeffreys, D. G., 1985: *The Survey of Memphis*, Egyptian Exploration Society.
Jondet, G., 1916: *Les ports submergés de l'ancienne île de Pharos*, Institut Egyptien vol. IX, Le Caire.

Kerisel, J., 1973: *Bicentenaire de l'essai de Charles Augustin Coulomb, présenté à l'Académie Royale en 1773*, Introduction. Presses de l'Ecole Nationale des Ponts et Chaussées, Paris.
--, 1975: "Old structures in relation to soil conditions", *15th Rankine Lecture, Geotechnique*, 251, n° 3, London, pp. 433–483.
--, 1982: "Les chantiers et monuments du passé: leçons pour l'ingénieur et l'architecte, Panama", *Annales de l'Institut Technique du Bâtiment et des Travaux Publics*, 410, pp. 37–54.
--, 1985: The History of Geotechnical Engineering up until 1700. *Golden Jubilee Book, XIth International Congress of Soil Mechanics, San Francisco*, Balkema, Rotterdam.
--, 1991: *La Pyramide à travers les âges*, Presses de l'Ecole Nationale des Ponts et Chaussées, Paris.
--, 1987 and 1991: *Down to Earth, Foundations Past and Present: The Invisible Art of the Builder*, Balkema, Rotterdam.
--, 1996 and 2001: *Génie et démesure d'un pharaon: Khéops*, Stock, Paris.
--, 2002: "The tomb of Cheops and the testimony of Herodotus", *Discussions in Egyptology 53*, Oxford, pp. 47–55.
--, 2003: "Le tombeau de Khéops et une vérification peu coûteuse", *Revue Travaux*, n° 796 April 2003, Paris, pp. 68–76.
Koldewey, R, 1913: *Das wieder erstehende Babylon*, J.C. Hinrichs, Leipzig. Translated into English by Agnes S. Johns, 1914: *The Excavations at Babylon*, London.
Lagerlöf, S., 1912: *The Wonderful Adventures of Nils.*
Las Casas, Bartolomé de, 1552: *Très brève relation de la destruction des Indes, 1992*, La Découverte, Paris.
Laronde, A., 1981: *Variations du niveau de la mer sur les côtes de la Cyrénaïque. Ports et villes englouties.*
Lagrange, M. J., 1905: *Eclaircissement sur la méthode historique.*
--, 1934-36: *Introduction à l'étude du Nouveau Testament.*
Laugier, Abbé, 1760: *Eloge de l'architecture du Panthéon.*
Lehner, M., 1974: *The Egyptian Heritage Based on Edgar Cayce Readings*, A. R. E. Press, Virginia.
--, 1983: "Some Observations on the Layout of the Khufu and Khafra Pyramids", *Journal of the American Research Center in Egypt* (JARCE) XX, pp. 7–21, Cairo.
--, 1985: *The development of the Giza Necropolis : the Khufu Project*, AfO 32.
Leroi-Gourhan, A., 1964: *Les Religions de la préhistoire*, Presses Universitaires de France, Paris.
--, 1965: *Le geste et la parole*, A. Michel, Paris.
Mc Cullough, D., 1977: *The path between the seas. The creation of the Panama Canal (1870-1914)*, Simon and Schuster, New York.
Málek, J., 1986: *In the Shadow of the Pyramids*, Orbis, London.
Maspéro, G., 1895: *Histoire ancienne des Peuples de l'Orient classique*, Hachette, Paris.
Montaigne, Michel Eyquem de, 1580-81: *Journal du voyage de Michel de Montaigne en Italie par la Suisse et l'Allemagne en 1580 et 1581.*
Mooney, J., 1893: *The Ghost-Dance religion and the Sioux outbreak of 1890*, Annual Report of the Bureau of American Ethnology, XIV.2, Washington.
Pagnol, M., 2001: *La gloire de mon père*, Fallois, Paris.
Patte, P., 1770, *Mémoire sur la construction de la coupole projetée pour couronner la nouvelle église de Sainte-Geneviève à Paris,* F. Gueffier, Amsterdam, Paris.
Paz, O., 1991: *Les Pierres lisibles*, rencontres internationales en l'honneur de Roger Caillois, UNESCO, Paris.
Peacock, J., 1855: *Life of Thomas Young*, J. Murray, London.
Penone, G, cité en 1989: Celant, G., *Giuseppe Penone*, translated by A. Machet, Electa-L. & M. Durand-Dessert, Milan-Paris.
Perring, J. S., 1839-42: *The Pyramids of Gizeh: from actual survey and admeasurement*, J.Fraser, London.
Perronet, J. R.: *Description des projets et de la construction des ponts de Neuilly, Orléans et autres.*
Petrie, Sir W. M. Flinders, 1883: *The pyramids and temples of Gizeh*, Field & Tuer, Londres.
Proust, M., 1954: *A la recherche du temps perdu*, La Pléiade (Gallimard), Paris.
Ramond, S.: Musée de la mémoire des pierres, Verneuil-en-Halatte (Oise).
Renan, E., 1883: *Souvenirs d'enfance et de jeunesse*, Calmann-Lévy, Paris.
Renard, L., 1867: *Les phares*, Hachette, Paris, republished in 1990, L'Ancre de Marine, Saint Malo.
Rondelet, J.-B., 1797: *Mémoire historique sur le dôme du Panthéon*, Du Pont, Paris.
Schmiedt, G. G., 1981: "Les viviers romains de la côte tyrrhénienne", *Histoire et Archéologie*, 50.
Séjourné, P., 1913-16: *Grandes voûtes*, 3 vols., Bourges
Siegfried, A, 1940: *Suez, Panama et les routes maritimes mondiales*, A. Colin, Paris.
Smeaton, J., 1791: *A narrative of the building and a description of the construction of the Eddystone Lighthouse with stone*, G. Nicol, London.
Taine, H. A., 1866: *Voyage en Italie*, Hachette, Paris.
Viollet-le-Duc, E. E., 1866: *Dictionnaire raisonné de l'architecture française du XIe au XVIe siècle*, A. Morel, Paris.
Vyse, R. W. Howard, 1840-42: *Operations carried out at the pyramids of Gizeh in 1837*, J. Fraser, London.
Wilkinson, R. H., 2000: *Temples of ancient Egypt*, Thames & Hudson, London.

Illustration credits

Artheme Fayard: Figures 14, 16 and 17.
Brigham Young University (USA): Figure 23.
© Compoint (Stéphane): Figure 25.
Dagli Orti: Figure 45.
Ecole Nationale Supérieure des Beaux-Arts, Paris: Figure 7.
French Ministry of Culture and Communication, DRAC of the Rhone-Alps region, regional department of archaeology: Figure 18
Gamma: Figures 8 (© Marc Deville) and 24 (© Brissaud/Saola).
Gemäldegalerie (Dresden): Figure 72.
GEMOB-Bonnet-Laborderie: Figures 68, 69 (© Emile Rousset) and 70.
Getty Images: Figure 9 (© Ed. Freeman).
© Franck Goddio/Hilti Foundation: Figures 96, 97 and 98 (photo Christoph Gerigk).
Hoa-Qui: Figures 15 (© F. Gothier) and 37 (© Patrick de Wilde).
Magnum: 46 (© E. Lessing).
Pour la Science: Figure 57.
Rapho: Figures 3 (© Gérard Sioen) and 5 (© Georg Gerster).
RMN (Louvre Museum): Figures 20 (© Daniel Lebée), 21 (© Franck Raux), 47 (© Gérard Blot), 48 (© Franck Raux).
Scala Group (Firenze): Figure 73.
Sipa Press: Figure 35 (© Gregorio Borgia).
Taylor & Francis/Balkema: Figures 6, 38, 39, 40, 41, 55, 60, 65, 66, 71, 74, 81, 82, 83, 86, 87, 88, 92.
Photographs and documents of the author: Figures 1, 2, 4, 10, 13, 27, 28, 42, 44, 50, 52, 53, 54, 89, 100, 101, 102, 103, 104, 105, 106.
Photographs free of right: Figures 11, 12, 33, 34, 43, 49, 51, 59, 61, 62, 63, 64, 65, 66, 67, 75, 76, 77, 78, 79, 80, 84.
Photographs rights reserved: Figures 22, 26, 29, 30, 31, 32, 36, 56, 58, 85, 90, 91, 93, 94, 95, 99.

Index